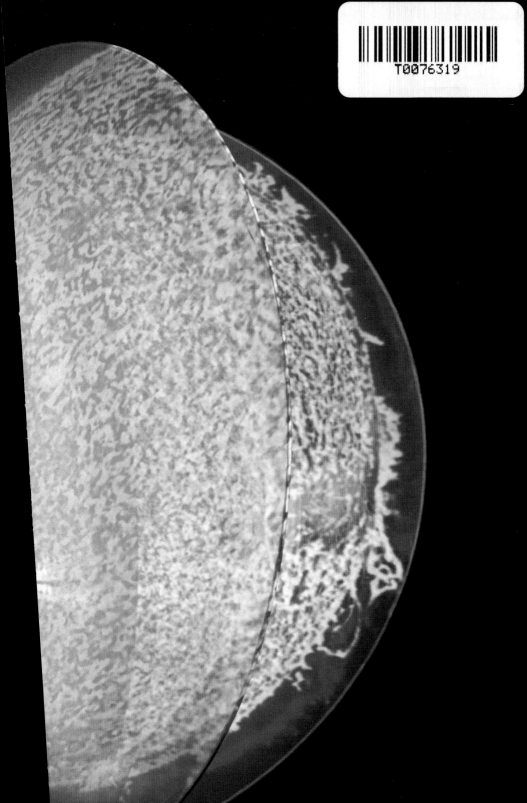

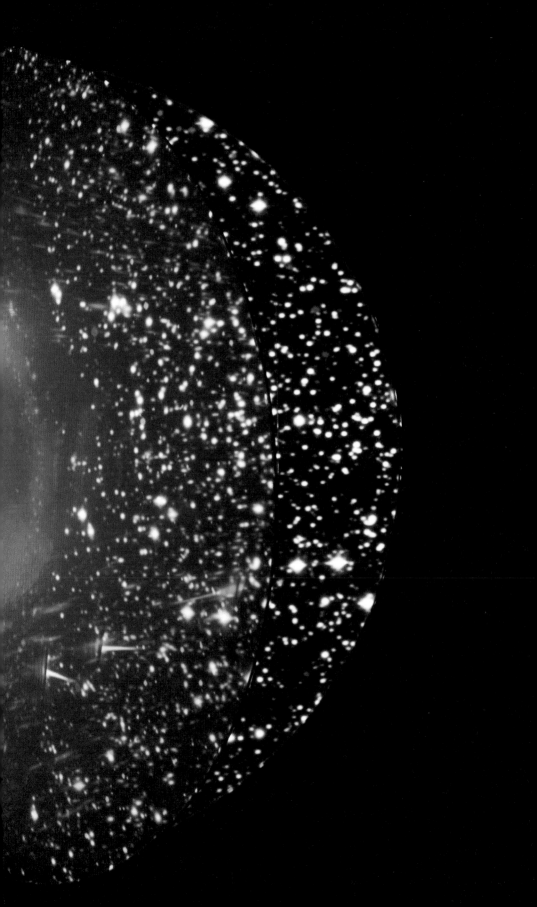

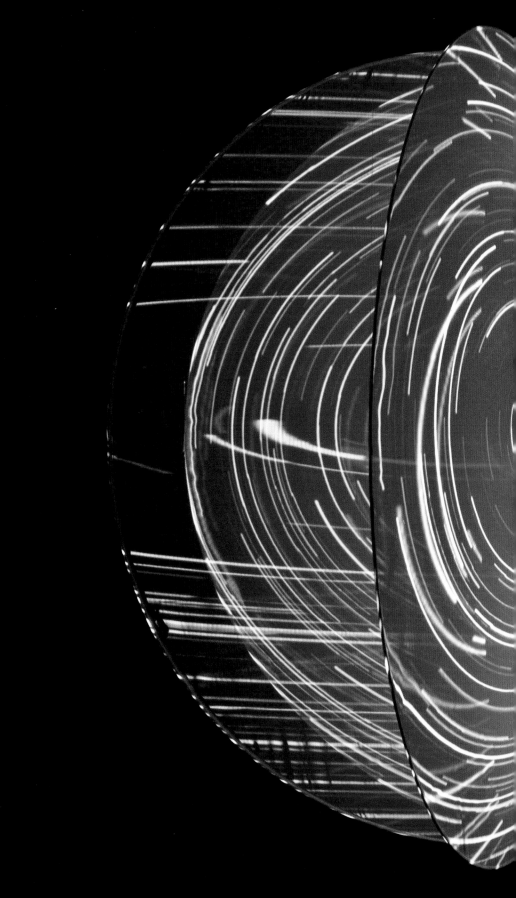

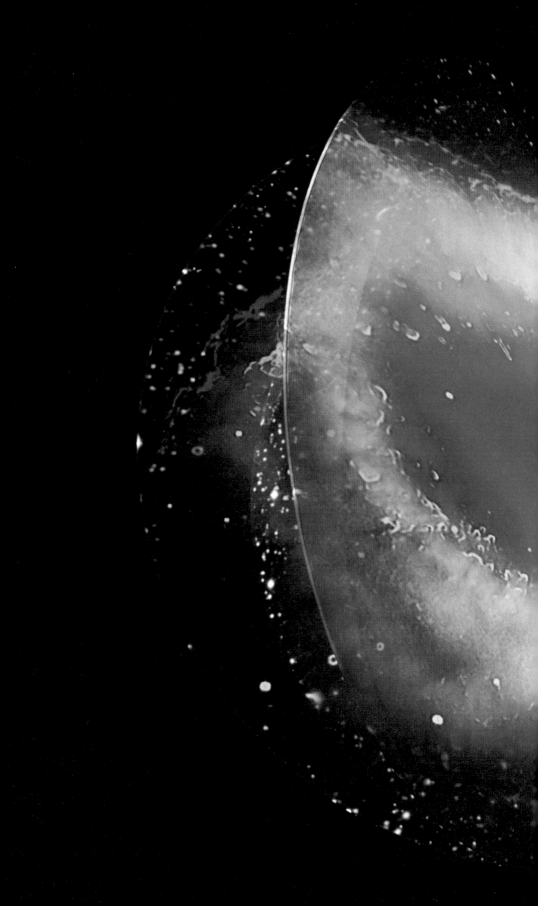

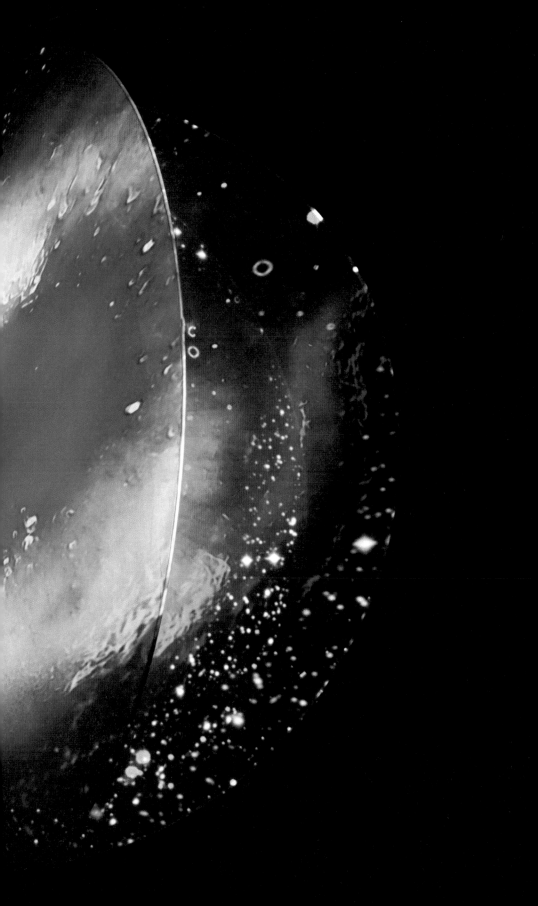

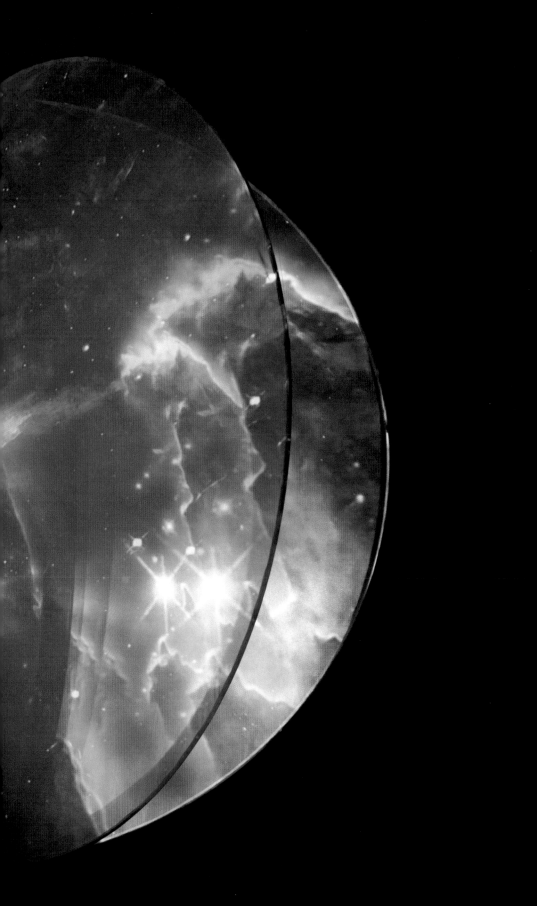

You've just travelled through the Universe. Or, at least, the interpretation of it by the artist Eugènia Balcells in an installation that was born at the Roque de los Muchachos astronomical observatory in the Canary Islands. There, the artist took stock of current astrophysical knowledge and translated it into her piece *Universe*. A vast amount of knowledge that we have gleaned based on the light that comes to us from the far reaches of space.

Thanks to light, we know that the known universe is made up of just a few elements. There are only 92 substances that combine to form everything we see, just like the letters that make up the words used to construct immortal verses, or to refer to things that exist only in the human mind. Like the pieces of an alphabet of life, the elements collide and blend together to forge wonderful combinations: a tiger, a redwood forest, a star, a human being.

According to Eugènia Balcells, light is the voice of matter. If we are all made of matter, the light from the elements is also our voice.

My Name Is Univers

Toni Pou

Edited by
Eulàlia Bosch

Prologue

Toni Pou

I came across *Homage to the Elements* by chance while doing research for a feature on the International Year of the Periodic Table, celebrated in 2019. Although it immediately caught my attention, my article didn't mention the mural by Eugènia Balcells: a periodic table in which, in place of the symbols of the elements and their properties, the cells are filled with the colors of light that each element emits in a unique way – in other words, their personal and non-transferable light footprint. In the article I did explain, among other things, that when Dmitry Mendeleev was head of the Russian Bureau of Weights and Measures, he analyzed vodka and concluded that if the proportion of alcohol was 38 percent by volume, the amount of heat produced when it was consumed would maximize its flavor. I also cited the largest periodic table in the world, which occupies 150 square meters of the Faculty of Chemistry at the University of Murcia, and the smallest one – engraved with an electron microscope on a hair belonging to the British chemist Sir Martyn Poliakoff. I also wrote about more exotic periodic tables, like the ones used to classify jams or styles of beer, football players, art movements, writers, or clothing collections, but I didn't mention anything about *Homage to the Elements*. I felt like, if I was going to say anything, even briefly, I needed to know more about it. I suspected that this seemingly simple work was hiding something that was striving toward a totality. And given that impression, my gut was telling me I should take my time with it and approach it outside the hustle and bustle everyday life. But, of course, the thing about the hustle and bustle of everyday life is that is goes on every day.

A few weeks later, as the daily frenzy continued as deafening as ever, I got an email from Eulàlia Bosch inviting me to have a conversation about *Homage to the Elements*. On the phone, she told me about the book you're holding in your hands, which was just a project at the time, an idea, and that's when I started delving deeper into the piece. I immediately realized that it does, indeed, contain a spirit of universality that can be read in many different ways, but which always digs down to the root of things, of matter, of color, of movement, of creation, of science. As I was talking to the artist a few days later, I became convinced that something had to be done to communicate (at least) part of all the things that can be found in this mural.

Having spoken with people from all over the world and from a wide variety of cultural and scientific backgrounds, my perspective on the work has, naturally, been enriched. And yet I find that it still has a patina of mystery that gnaws at me. I would say this happens to me with any creation — be it scientific, musical, literary, etc. — that captivates me, and which I've explored in depth. And I would dare to say that it isn't attributable to the works themselves but to a certain aspect of human nature. How can we be so fascinated by something that is, apparently, useless? The German historian and philosopher Oswald Spengler said that humans are the only animals that understand death. Other animals grow old, yes, but their awareness is limited to the present. Humans, on the other hand, fear death, and that fear lies at the root of religious, scientific, philosophical, or artistic inquiry because we're hoping to find some semblance of permanence in that search. The physicist Brian Greene has said that we are the product of eternal laws and that, in spite of everything, we exist for just a brief instant; that we're governed by laws that are blind to fate, but we never cease to wonder where life is taking us; that we are shaped by laws that don't seem to respond to any reason, yet we insist on searching for purpose and for meaning. And all these things – which pursue so intently because we're aware of our own mortality, and which the writer Franz Kafka identified as the need for something indestructible – we seek them out today largely in art and science, because somehow they offer a hint of eternity, a glimpse into the infiniteness of time, into immortality. After all, every football that has ever been or ever will be kicked moves in accordance with Newton's laws. That has always been the case, and there's no reason to think it will ever be different in the future.

Science, in contrast, and some forms of art, can arise from a more mundane evolutionary usefulness. Scientific research consists in discovering patterns in natural phenomena, a skill that comes in handy for understanding the best time to harvest certain fruits, for predicting where to find the animals that serve as a source of protein, shelter, and materials for making tools, or for knowing when the river will dry up and you'll need to settle in a different valley. On the other hand, an appreciation for music requires an acoustic discernment that can be useful in detecting hazards or finding food sources in a natural environment. In turn, the ability to make up stories can be harnessed to imagine situations that haven't occurred yet and be prepared for them. Plus, as Greene also points out, reading about the hardships of Don Quixote gives us an understanding of human nature that we would never find in a description of the movements of the atoms that made up the famous knight-errant.

In that sense, how should we understand the visual or abstract arts, like Eugènia Balcells' work, and to which *Homage to the Elements* belongs? Do they have a pragmatic origin, or are they merely (!) trying to capture eternity? An evolutionary explanation, like the one given for music or literature, would argue that artistic creation and appreciation sharpen ingenuity, they improve creativity and foster a taste for innovation, for unrestricted symbolic thinking, and collaboration. On the other hand, the writer Saul Bellow said that "only art penetrates what pride, passion, intelligence and habit erect on all sides – the seeming realities of this world. There is another reality, the genuine one, which we lose sight of. This other reality is always sending us hints, which without art, we can't receive."

Ultimately, though, what all this disquisition does is look for answers to a question that may not have any, and it proves that in art, as in science – no matter how much it enriches us and encourages us to live life in new ways – there is always a hint of mystery. A mystery that may be impregnable and definitive. Because the limitations of the human brain are obvious: it is an organ limited in size by a relatively small skull, and which developed in response to the selective pressures of an environment very much like the savannah. As a result, we may not be able to unravel those kinds of mysteries. That said, simply identifying and formulating them is a passport to that something indestructible Kafka wrote about.

And *Homage to the Elements*? Is it catching hints from Bellow's genuine reality? Is it a ticket to something indestructible? Does it hone our ingenuity, creativity or innovation? Does it encourage symbolic thinking and collaboration? One possible answer – however, incomplete, inaccurate, tempting, but alive, penetrating, and honest – is this book.

Capturing the real

Jordi Balló

When I read *The Immortal Seed*, years ago, I discovered a synthesis in it, which might be described as immortal as well, since it has come back to me every time I'm watching a TV show or a film. The strangest thing about it, though, is that this accompaniment isn't something voluntary. I would say it isn't entirely conscious either. It just happens. And it means that Jordi Balló and Xavier Pérez were successful in what they set out to do with their book. They say that there are a series of universal stories, created in literature in works ranging from the Bible and *The Odyssey* to *Madame Bovary*, that audiovisual narratives adopt and reformulate over and over. In *The Searchers*, John Wayne is Ulysses. E.T. is the messiah. And, as a result, we can establish a phylogeny for nearly any audiovisual work. The synthesis always comes from finding a simple idea that can be used to explain many different things. In this case, the plot of a literary work can be used to explain many films. The idea, ultimately, is to look for patterns, to uncover the commonalities between different things. And that act, which seems so simple, but which requires a huge amount of knowledge, in addition to its undeniable scientific connotation, is precisely what Jorge Wagensberg defines as the process of understanding. A process which chance and time very likely selected to be pleasurable because of the innumerable benefits it has provided throughout the history of human evolution. Imagine how useful it was to find patterns in the movements of animals 100,000 years ago. As is clear in all his work, Jordi Balló has the ability to weed out the details and the noise of contemporary audiovisual production and see through to the patterns. His is a perspective that understands and generates understanding. If you want to deepen your enjoyment of the shows we all watch as we're slumped on our sofas, seek him out. Listen to him. Read him.

How did you find out about Homage to the Elements?

Years ago, I had followed Eugènia Balcells during the period when she was making conceptual films, and I identified closely with her reflection on the images being generated by films and the media. I kept an eye on her work, and she did a piece on Virginia Woolf, *A Space of One's Own*, which was based on an idea that has a lot in common with her latest work. In that piece, she was already playing with the idea of light and transparencies, a very different approach from the one she had taken initially, when she was working more on a counterimage of the media. In contrast, in this piece I was very surprised by the expressive cultivation of light and transparency, and the treatment of domestic spaces. At the time, I was also really interested in her piece *In the Heart of Things*, in which she depicted different rooms in her house, because it reminded me a lot of Chantal Akerman's first film, *Jeanne Dielman, 23, quai du Commerce, 1080 Bruxelles*, the story of a woman going about her life inside her home. It's a film that plays with time and a nearly real-time rhythm, and which moves through domestic space with an unexpected upsurge in the level of drama. Those pieces made a real impression, and then I saw *Homage to the Elements* at Arts Santa Mònica. It's a piece that I can't really separate from *Frequencies*, because I remember them, together, as a captivating installation. I imagine I'm not the only one who sees them as a diptych that plays with the same idea using different materials.

Balcells has worked with very cinematic elements that are typical of the audiovisual world. Where do you see her fitting into that world?

I see her as part of the shift toward decontextualizing the images provided by film, a movement that carried a lot of weight at a certain moment in time. For example, I still think her piece *Boy Meets Girl* was a brilliant idea, because it decontextualizes the images by associating them with the apparently random nature of a slot machine. It was a fantastic idea and, plus, it was almost a prelude to the whole concept, which I've also studied, of how visual motifs are reproduced at different scales in the world of film and in the art world, so that they call up this sort of iconographic memory we all carry with us. I think that's a really fruitful avenue in her work, which is also very fragile at the same time.

> She's also considered an experimental filmmaker,
> isn't she?

There are a series of experimental filmmakers who have truly been
able to portray the idea of a cosmogony, a universe, of the landscape
understood as a universe. And that's something very different from
what happens in a Western, for example, which focuses on a very
specific landscape where the action takes place. But who has been able
to film a landscape and, through that landscape, imagine the planet,
the world and its place in the universe? Funnily enough, it has been
that group of experimental filmmakers, who have largely been ignored
in film history until now, who were able to do that. In their fragility,
they were able to present that idea of the landscape as a universe. I'm
thinking of Jonas Mekas who made *Walden*, or of Ben Rivers filming
in the desert for *The Sky Trembles and the Earth is Afraid and the Two
Eyes Are Not Brothers*, or of how Michael Snow filmed *La Région
Centrale*. I've always understood Balcells as part of that universe,
especially in this latest period, where she has been working with the
idea of light as a fundamental element. Through fragility, an image
can be given that cosmogonic capacity. Because of the forcefulness of
major studio films – which are fantastic, don't get me wrong – they
convey a visible world in which it is incredibly difficult to see any
capacity for expansion. Whereas experimental filmmakers like Jonas
Mekas present clear examples of that cosmogonic capacity. It would
be impossible to write a history of film today without situating him
as one of the most important filmmakers, but for years he was seen as
someone who merely compulsively filmed the world that existed ten
feet around his personality.

> That current originating in experimental film has
> increasingly permeated the film industry, and that
> has translated into a rise in the documentary genre.
> It has also tied in with the ascent of the literary or
> narrative journalism practiced by authors like Leila
> Guerriero o Martín Caparrós, which originally
> began in the United States in the 1960s.

The rise of the documentary format is undeniable, but interestingly
enough, it wasn't spearheaded by journalists. That could have been

the case, but it wasn't. The boom started when filmmakers who had usually worked in other genres, and who might even have felt that the term *documentary* wasn't an advantage for them, discovered that it offered a huge amount of freedom, where chance, unpredictability, experimentation were all accepted. It's a realm that demands something which, logically, traditional television can't offer. And that reclaiming of the word *documentary*, divorcing it from the realm of journalism, led to an international explosion. Suddenly, the best filmmakers were able to see in the documentary format what Agnès Varda used to suggest, "Since women have come late into directing, we need to leave traces of ourselves in the films we make, so that the images aren't colonized by the male perspective." I think Eugènia Balcells, Chantal Akerman, Agnès Varda, Naomi Kawase, and many other women have worked a lot with this idea that universal stories can be constructed through the self.

At first, it seemed like it was a space only for experimental filmmakers, but the genre has been on the rise and we've now come into a very interesting moment. Especially because the history of film has had to atone for the fact that no one had ever included artists like Balcells in the genealogy of cinema. They were considered video artists who showed their work in museums, which is great, but this whole process has also been important in making it possible for those artists to feel like a part of the larger cinematographic tradition. And as a result, all films, theirs and all the others, come into contact, and different things are demanded of them.

> When you talk about that space of freedom in documentaries, which is so interesting, do you mean that in other films, in fictional films, so to speak, it doesn't exist?

Since the year 2000, the documentary format has been gaining ground internationally, and in a way, what that means is that it has the language to explain what's going on and to satisfy a desire for the real on the part of both filmmakers and audiences. And what is the real? It's our conceived reality, our constructed reality. A documentary has no qualms about recognizing that, from the outset. As a result, fiction has also started along the same path, because it was beginning to see too much of a contrast. The strongest fictional experiences from recent years are films that have done what Agnès Varda affirmed about leaving traces of oneself. In those films, the actors leave a part of themselves. Today, when it comes to the fictions that interest us the most, they affect us because we realize that they aren't prefabricated operations with a pre-defined script, where the actors are only responsible for embodying what a screenwriter has imagined. The film *Marriage Story*, for example, directed by Noah Baumbach, is the story of a couple in crisis, but it's clear that a lot of it is based on contributions from the actors. That's why a film like Stanley Kubrick's last, *Eyes Wide Shut*, improves so much with time. Kubrick had the great idea for Tom Cruise and Nicole Kidman to star in the film because they were a couple that was heading for a separation. And, to some extent, you can't watch *Eyes Wide Shut* without being aware of that level of tension.

> That same influence is also filtering into television series.

We've come into a very interesting time, populated by various different tensions: on the one hand, between experimental film and documentary film; and, on the other, between documentary film and fiction, and, of course, also between experimental film and pure fiction. That last tension is represented in an incredible episode of *Twin Peaks* (episode eight, season three) by David Lynch. The episode is an absolute tribute to experimental filmmakers. At one point in the episode, there's an image that lasts almost two minutes, as the camera enters a room that is just pure light. David Lynch's light is like the light inside an atomic bomb. The particle of light is hell. No one had ever filmed an atomic bomb understood in that way. It's an utterly abstract two-minute shot – which is a really long time in television – and it's just light and particles. The scene would clearly be the flip side of Balcells' work, and now I'm thinking of *Frequencies*. One is hell, and the other is a cosmogonic and vitalist idea understood from the aesthetics of her own body, her gender. Here David Lynch pays a clear tribute to experimental filmmakers, and to Kubrick, and Cronenberg and other directors. But in any case, the clip attests to a fantastic moment, in which someone who has a lot of money to make a TV show takes a tour through the expressive world of the filmmakers of fragility.

Does the idea of the documentary emerge in parallel to the birth of film?

Ultimately, everything comes down to a three-month period. When the Lumière brothers had finished filming their famous scene of workers exiting the factory, they could have figured that they had the entire world to capture on film and set out to do just that. But instead of moving on, they filmed the same thing again, twice. So there are three versions of the factory exit, filmed with a few months' difference. Why did they film the same thing three times, when there was a whole world out there to capture? Because the first and second time, the shot didn't include the factory door closing. There was a horse and cart that was slowing it down. Since they only had 47 or 48 seconds of film, the third time they got rid of the horse and cart, which gave the doorman enough time to shut the factory door. Obviously, the piece, although its nature is documentary, was staged cinematographically. That's the difference. The documentary format has been around since the very beginning of cinema, that's key, but it has always been permeated by this idea of construction. As I see it, the Lumières' third version is even more of a documentary than the first two. At any rate, documentary film has gone down the road toward that third version. The brothers Lumière said that, in order for a film to be good, you had to have enough time to shut the door.

On the other hand, I'm under the impression that this tension, already felt by the Lumière brothers, has permeated the entire history of film. And over time, the viewers' gaze has also changed, and they have been able to recognize powerful, undeniable things in its fragility. Documentaries have never claimed to be divorced from fiction or construction; if that were case they would no longer merit the name. And as the Nouvelle Vague and other movements from the 1960s proclaimed, the best fiction has always incorporated this idea of introducing elements of chance into the process of filming.

> The boundaries are fuzzy, as is the case with
> everything.

Exactly. And it was the experimental filmmakers who moved away from breaking down those boundaries because they figured they were already getting recognition from museums and other more artistic spheres. But the field of documentaries kept at it. The filmmakers who have come into documentary filmmaking haven't given up. And that has meant that people were able to see *Work in Progress* at a multiplex, with a Hollywood movie on the screen next door, and they were forced to decide whether to spend their money on one or the other.

> In that sense, do you think that *Homage to the
> Elements*, and *Frequencies* as well, have something of
> the documentary in them?

If we set them up side by side, the images of *Frequencies* and the scene from *Twin Peaks* I mentioned earlier respond directly to one another. Both of them are pure light. One is hell and the other is a cosmogonic idea. In that sense, *Frequencies* is realistic. And, in the same sense, *Homage to the Elements* is too, of course. And that, for me, is the great leap forward in this double play of elements and light. That's real, it's the real. It isn't an abstract conceptual contrivance, it's the real in its pure state. In the same way, David Lynch's image is also the real. David Lynch has his demons, and no doubt living in David Lynch's mind would be a bit terrifying. We'd be better off in Balcells' mind. But, ultimately, they are both ways of describing the cosmos.

> Balcells' work is based on scientific ideas, but there
> aren't many documentaries on scientific topics.
> When we talk about scientific documentaries, people
> think about animals or nature documentaries, which
> are a bit anachronistic.

I agree, but documentaries aren't ever about something. *About* isn't the right preposition for a documentary. Because when you say it's "about something", you're implying that the thing is pre-existing and the documentary explains it. Going back to the comparison between *Frequencies* and that clip from *Twin Peaks*, both of them focus on an element that, in theory, is only of interest to the scientific community.

But in this case, the artists take a closer look, they play with the weapons in their arsenal, they reveal something, and on the other side (that of science) they suddenly discover something. In that sense there has been a constant common ground, but there hasn't been enough communication to weave it together. For example, when we realize that non-linear narratives fit in so well with certain scientific principles, it's clear that having established a closer relationship earlier on would have been mutually beneficial. That relationship wasn't there, and a lot of its potential has been lost. Now, however, I would say that the exchange is picking up. I believe that there are scientists who think that certain works of art, and this piece by Balcells is a clear example, reveal things that the artist would probably never have thought could be of interest to a scientist. Someone might think that this view is a bit forced to fit in with the spirit of our times, but I truly think that there is a possible dialogue. And it has to do not only with scientific questions becoming the subject of documentaries, but also with the knowledge that scientific pursuits and findings can enter into dialogue with other disciplines. When Akira Kurosawa made *Rashomon* based a text by a Japanese writer who told a story from different points of view, semiologists and other experts had their say. *Rashomon* directly harnesses the scientific idea that observation changes the meaning of a phenomenon, but the scientific world didn't claim it as its own. It's a very typical case. Narratology claimed it, but without acknowledging the scientific explanation.

> If we can't say a documentary is *about* something,
> how should we talk about them?

It's not that we have to talk about them in one particular way... The point is that we call documentaries documentaries because they offer a degree of freedom and because they recognize that reality is a construct. Just like in the case of fiction, we call it fiction because we understand that there is a mechanism of forethought that isn't there in a documentary. A documentary is an attitude, a method, a way of approaching the real, a way of experimenting with forms; it isn't a genre waiting around for subjects to come around so it can explain them. The subject matter doesn't gain anything (although perhaps in some cases there is an opportunity for visibility that wasn't there before), nor does the documentary. Documentaries, like all types of film — and this isn't the case for other arts, but it is for film — aspires to centrality. That means that if a movie is about something, it is that thing. It may not aspire to be the whole thing, but its aim is to occupy the center – regardless of its budget or how many people will be able to see it. If the film is a great film, it will last, and everyone will be grateful for it. I have strongly defended the idea that a documentary is a mode of expression that has offered freedom to its authors and has been inclusive, and audiences have responded to it precisely because it has a degree of unpredictability. If what we want from a film is for it to surprise us, the best documentaries have surprised us. And the strength of documentary cinema doesn't come from the range of its subjects. For example, women filmmakers have made waves with the idea of how their own experiences can be included in a film. Some men have done it too, but women have done it across the board. Interestingly, some men have ventured into this introspection at the end of their careers, after making commercial films. But women have done it from the very beginning. And what kind of format was willing to accept those poetics of the self? The typical fictional film of the past sure wasn't. But experimental film could. And Balcells was one of the leading figures who pursued that. Documentary film has proven to be receptive to the poetics of the self.

Is that similar to what is called "autofiction"
in literature?
It's related. The film critic Mirito Torreiro says that the first autobiographical film in Spanish cinema is *Monkeys like Becky* by Joaquim Jordà, from 1998. We should be careful with statements like that, but upon close examination, you'll find that Víctor Erice, for example, never made any autobiographical films. *Dream of Light* isn't autobiographical. Jordà was the first, when he was making a film about lobotomies. Suddenly he had a stroke that left behind terrible consequences, but it situated him on the other side. In other words, he began making the film from the point of view of someone whose brain had been affected. That was a fundamental shift. That film opened a lot of people's eyes, in Spain, in Catalonia and abroad, because from that moment on, you could do anything you wanted.

Autofiction, on the other hand, has a bad reputation
in the literary world.

Because it has become a label. In that sense, when people started
talking about the boom in documentaries, we were all pretty wary, and
we kept our distance. The documentary isn't a stamp that indicates
an autofiction, it's a tool; the documentary isn't autofiction because
autofiction is a subgenre within a genre. The documentary is a way of
relating to the real. The documentary movement was very interesting
when it began around 1999. But in recent years the most interesting
movement that has taken place in film in Catalonia has been the
emergence of women filmmakers who put traces of themselves into
their films, like Agnès Varda said. *Estiu del 1993*, by Carla Simón, may
be autofiction, but it's a great film that would never have been made
without the documentary tradition. And yet the film's style is entirely
that of fiction, but it's inscribed within the work of people who had
already constructed a poetics out of their daily lives.

When it comes to portraying a difficult and
profound personal experience, like in the case of
Monkeys like Becky, fiction can't do it better.

No, of course. That's true. In fact, Jordà started out making a fictional
film about António Egas Moniz, the Portuguese Nobel laureate in
medicine who performed lobotomies on humans. But then he had a
stroke, and he turned it into something else. And in that indefinable
'something else' there is a part that is a classic documentary, a part
that is the reconstruction of an imaginary play staged by the patients
themselves, which takes place in a very real register, with an immense
emotion of reality, and then there is his shift to the other side after
the stroke. The work has three films within it, but it is an example
that shows how things can flow without being perfect. Now that
Balcells has finally made *Letters from Akyab* I feel reassured, because
I always told her that everything we were setting in motion with the
documentary format was for people like her. Balcells' work is held
in high esteem within the movement because it has a conceptual
dimension, but also because it has an intimacy that corresponds
perfectly with the register in documentary film that has emerged in
recent years.

Exploring Darkness

Priyamvada Natarajan

"From now on, I only want to talk to scientists," said Salvador Dalí at a certain point in his life. Science had always interested him, because it is a process that makes the invisible visible; it is an intellectual mechanism that brings about some kind of revelation. For Priyamvada Natarajan, the pursuit of science is tantamount to making a map that lays out everything we know, but which also leaves room for the unknown. Thanks to the 32 volumes of the *Encyclopedia Britannica* at her home in Delhi, when she was little she fell in love with maps, of both the Earth and the sky. And in every map, there is always a space for dreams and imagination. Perhaps that is why she now makes maps that attempt to make visible one of the most notable invisible things of our times: dark matter. Perhaps that is why, in addition to being an astrophysicist, she is also an avid and discerning reader of poetry. Following one of her lectures in Barcelona, as we ate at a delicious vegetarian restaurant, she confessed that her favorite map of the sky is the *Catalan Atlas* – dating from 1375 and attributed to the Mallorcan Jew Cresques Abraham – because it is the first map in which divine images are replaced by astrolabes and other instruments. A shift from the divine realm, imagined but invisible, to the measurement of the visible world. From myth and mystery to science. And with her characteristic insight and provocativeness, the reason for her inclusion in this book, she conjectured, speaking more softly now, whether, as suggested by the desire to make things visible in that map, the Renaissance might not have started in Al-Andalus as opposed to Italy.

How did your passion for maps begin?

Since I was very young I've been interested in the idea of exploration, the idea of going places that may be real or imaginary, that you can only get to by following a map. My interest in maps also has to do with the idea of making invisible things visible, and understanding visible things on a deeper level. In my career as a scientist, I have kept up that passion because I work making maps of the universe. One incredibly interesting aspect of maps, especially the older ones, is that they incorporate a kind of humility, by acknowledging that there are things we don't know. In the old maps used for exploration, there was always a clearly marked *terra incognita*. And that boundary between what we know and what we don't was something that fascinated me. Our goal as humans, and specifically scientists like me, is to push back that boundary, to expand our understanding. That desire to push my own mind, and our collective understanding, as far as possible is one of my main motivations. And, in my case, the inspiration came from my longstanding passion for maps.

Today we have powerful computers, artificial intelligence, complex algorithms, but when we're faced with a problem, it seems like we still need to represent the information visually, whether it's by sketching with a pencil and paper or using computers in a more sophisticated way. Ultimately, it boils down to making maps like the ones people have been making for centuries.

Because making maps isn't just drawing places; it's drawing the connections between those places. My research moves forward thanks to the ability to see unexpected connections, thanks to the ability to look at one place on a map and then look at a completely different place in the same way. The transfer of knowledge that takes place between those places on the map is very fruitful.

Is the periodic table a map?

It's a very powerful map because it is a way of showing all the things we know but, at the same time, it also shows what we don't know yet. I study dark matter, for example, which is an unknown and elusive type of matter that isn't on the periodic table. We know that it isn't made up of ordinary atoms and that it doesn't interact with light. That means it doesn't emit light in the form of a spectrum; it doesn't have the kind of signature that Eugènia Balcells used to construct her magnificent piece *Homage to the Elements*. By its nature, dark matter will never fit onto the periodic table; it can't be in a new column or an extra row. It's something entirely different. That said, the fact that it isn't on the periodic table doesn't mean that it isn't real. Although it doesn't emit light, and therefore we've never seen it like the other kinds of matter, we have proof of its existence.

> What are your impressions of *Homage to the Elements?*

In essence, what Balcells did with this piece was to create a map that distills the essence of all known matter, because the colors of the light emitted by each element are determined by the structure of the element's core and the number of electrons it contains. What I find most remarkable about the piece is that light really is the quintessential cosmic messenger, the bearer of all knowledge. I work with cosmic distances that we can't grasp intuitively and with objects that are so immense that we can't imagine them. And in all that enormity there is an emotion that emerges in the process of understanding the cosmos. *Homage to the Elements* captures all of that in a very powerful way because it shows us everything we've discovered so far and it tells us what isn't there, the *terra incognita* that has yet to be explored. I understand the piece as an invitation to explore further.

> There are so many things we don't understand, like dark matter, but there is also a lot that we understand pretty well.

We've discovered what things are made of, from a material standpoint. But what we see on the periodic table is only about 5% of all the energy in the universe. And in that fact, there is a tension between the significance and the cosmic insignificance inherent in the human condition. There are lots of things we have discovered, but there is still so much left to discover. I think it's very impressive that we've discovered everything that's on the periodic table. Plus, we know that 98% of all the matter in the universe is helium and hydrogen.

And eight more elements – oxygen, carbon, iron, neon, nickel, silicon, magnesium and sulfur – account for 1.96% of all matter. The other elements make up only 0.04% of all matter. Another interesting thing is that helium, hydrogen, and lithium were formed during the first three minutes of the universe. The remaining dozens of elements on the periodic table – which we have studied and understand so well today – were synthesized in the cores of stars or in the collisions between neutron stars. We're made of stardust, and Balcells' piece makes that very evident.

But how do you get from stardust to human beings? It's a process that we're slowly beginning to understand. Every day we know more about how those atoms came together to form our cells, our organism as a whole, with all its senses and intellectual, emotional, and psychological abilities. But there is one very important thing that we still don't understand: the nature of our mind and consciousness. We know that we are more than simply matter because we are mind and matter. Although we're just a collection of atoms, we are much more than a lump of matter. In that sense, I see a unifying principle in Balcells' work. It doesn't matter what we look like, what language we speak, what culture we come from, what kind of food we cook, or how we approach the world. Our eyes are sensitive to the same wavelengths, and we're capable of seeing the colors of the spectra of elements. And those signatures are unchanging, whether you measure them on the Earth or on another planet. For us, having discovered something that is so fundamentally immutable, which also explains what we're made of, is remarkable. And *Homage to the Elements* is a very clever reminder of everything that we know and everything that is still left to find out. We know that these elements are the fundamental building blocks of life, but we are missing the alchemy, the magic alchemy of how the elements join together to create human beings with a mind and an artistic drive.

You were saying that *Homage to the Elements* in an inspiration to explore outside the periodic table, to enter into that *terra incognita*. How do we know that dark matter is part of that *terra incognita*?

Light is the bearer of all the knowledge about the universe. We know that matter causes the path of light to curve, because there's a relationship between the presence of matter and the form of space. We might imagine space like a giant sheet of stretchable fabric, and matter makes potholes in it, so to speak. The light emitted by a far-off galaxy travels toward us through that sheet, which is full of potholes because the universe is full of matter. Since there is nothing outside the sheet of space, when light travels through it from a far-off galaxy, its path is altered, and that distorts our vision of the galaxy. When we analyze that distortion, we see that more than 90% of it can't be explained by visible matter alone. There are potholes in the sheet created by matter that we can't see, which we call dark matter, and based on that analysis we can calculate how much dark matter there is.

Is that enough to substantiate saying that the matter we can't see won't ever fit on the periodic table?
There is another important line of evidence. Although dark matter isn't made up of ordinary atoms and it isn't on the periodic table, we know that gravity is the only force it reacts to and generates. And gravity, in turn, causes movement. When we calculate the velocities of the stars in a galaxy, we find that they move much faster than what would be caused by the gravity of visible matter. In fact, if there were only this gravity, with the velocities that have been observed, the stars would shoot out of their galaxies, they wouldn't be able to hold together in a structure. There isn't enough gravitational glue in visible matter to keep them together: we need dark matter to explain their movements. Based on this and other evidence, we can take an inventory of the universe, and we see that, however you go about it, the results are the same. What is most fascinating of all is that the most important entities that shape and structure our universe are invisible, untouchable, and essentially unknown. But they're omnipresent.

How is the nature of dark matter being investigated?
That's the biggest challenge. One of the strange characteristics of dark matter is that it interacts with almost nothing, and that is a blessing and a curse at the same time. It's a blessing because, in truth, dark matter particles are travelling through us constantly and, luckily, they

don't destroy us or alter our bodies, and life is possible. On the other hand, because they interact so little, they are very difficult to detect. According to the standard particle physics model, there are a few particles that are candidates to make up dark matter. One of them is the neutralino, created in the primordial universe. Neutralinos are detected using liquid Xenon, a noble gas that has a crystalline structure in liquid form. If a dark matter particle were to pass through it, the crystalline structure would tremble, creating an earthquake of sorts. Unfortunately, that earthquake is very hard to detect, because there are many other factors that could cause it. Until now, that type of experiment has been unsuccessful. Another kind of particle that is a candidate for dark matter is the axion. Axions were also created in the primordial universe, and they have one very interesting property: they can decay into photons. This would be visible because there would be a little glimmer, a flash of light when those particles decay. The axion theory was proposed years ago, and unsuccessful experiments have been going on for a long time. But our theoretical understanding of axions has improved a lot in the last 10 or 15 years, and now there are six or seven new experiments that have just begun around the world.

Is that drive to explore the unknown, like dark matter, a common characteristic of art and science?
"Only mystery allows us to live," wrote Lorca. This curiosity of the mind in approaching mystery, what is invisible, and making it visible in some way is a drive that truly connects art and science. Things that are invisible may be real or imaginary. And those are the two ways that art and science approach exploration. Science is based on laws of nature that compel us to think in particular ways, and that's useful because nature is complex, and those laws give us a framework for thinking about what we don't know. Art doesn't have as many restrictions; it has a broader and freer range when it comes to thinking about imaginary things. And yet, art and science investigate the world and try to infuse it with meaning. There is a wonderful verse by Machado that says, "the truth is also invented." Art and science invent the truth in different ways, but ultimately they have a similar purpose. Using one language or another, they try to put into language what has been impossible to express up to that point. In a

way, science and art try to understand this unique and spectacular alchemy – this alchemy of order, of what is real and what is imagined, this transmutation of the imaginary into the real – and to understand how one becomes the other. Humans have always been fascinated by alchemy. And I think it's because alchemy appeals to a fundamental impulse, the drive for such deep understanding that it's possible to transform one thing into another.

Do you see a kind of alchemy in Balcells' work?
There's a popular belief that art is very emotional, irrational and psychological, whereas science is rational, logical, orderly and structured. And that art and science are very different activities, two drives that can't be reconciled. But that's a false dichotomy because science is incredibly creative; it's fueled by passion. And we also know that, on a more fundamental level, science isn't as objective as we think. Quantum mechanics tells us that the very act of observing a system alters it. That dichotomy comes from the separation between the observer and what is being observed. Scientific tradition assumed that there is a very clear division between what is being observed – the natural world – and the observer – us – and that an objective perspective is possible. But once you've been dedicated to science for some time, you see that the distinction between object and subject is pretty artificial. Science can't be that objective, because the very act of observing nature changes it; it has an impact on the systems. We aren't inert observers. And in that sense, in Balcells' work, there is a very clever gesture that consists of overturning that vision of art as an irrational and unstructured activity. *Homage to the Elements* is a work of art, the essence of which is order and structure. Plus, it is a profound order because the universe is made up of these fundamental elements that exist in a unique form with their spectral signatures. With this piece, Balcells challenges what art can be, and she demonstrates that science and art together can create new types of material objects and deeper understandings of the world.

Flying with Our Feet on the Ground

Amandine Beyer

Imagine the scene. We're in the auditorium of the Espai Ter in Torroella de Montgrí, in the Empordà. The concert began more than an hour ago. The violinist Amandine Beyer and her ensemble, Gli Incogniti, have played five Vivaldi concertos. They're getting ready to start the last piece, the concerto in C major, *Per la solennità di San Lorenzo*. The performance is called *Il mondo al rovescio* [The World Upside Down], because in these concertos, Vivaldi lets himself get carried away by fantasy and, in a playful spirit, he turns some of the musical elements that were considered untouchable on their head. Up until now, the music has flowed perfectly with the cheer and levity of the Baroque and a classical mise-en-scène: the musicians are dressed in black and Amandine Beyer, soloist and conductor, is wearing a turquoise outfit and high-heeled shoes. As she tunes her violin, she takes a few steps and it's clear – as it was when she walked on stage – that she isn't comfortable. She bends down to touch her ankle. She lifts one foot, shifts a leg. Finally, like Vivaldi, she changes one of the unwritten rules of classical music and she takes off her shoes. Whispering can be heard in the audience. She turns toward them and, chuckling, she says "I'm done with those." The last concerto begins, and it feels like the music is soaring even higher than before. It's clear that the high-heeled shoes, one of the most contrived objects in Western civilization, were constraining not only the violinist's feet but also the music itself. Maybe because, for someone like her, music comes from her whole body and not just her fingers. To let the music take flight and carry with it the audience's understanding, Amandine Beyer decided to break with protocol and play with her feet on the ground. The music, she has made clear, comes first. And the spectator, impassioned, can't help but wonder why she didn't do it earlier. And why every musician who walks onto a stage doesn't do the same.

Is the violin a difficult instrument?
It is, and it doesn't get easier!

When you play the piano for the first time, every
key makes the right sound. But the first time you
draw a bow across the string of a violin, it's a very
different story.
True. But all told, it's probably worse for everyone else than for the
person playing. For example, I started when I was quite young, and I
don't have any bad memories.

What has your relationship with the instrument
been like?
It's an intense relationship, a way of life. Music permeates your life,
and vice versa. But this last year has been strange, because for the first
time since I was seven years old – I'm 46 now – I decided to take a
break. I'd never done that before. I would say that I played nearly every
day of my life before that. I wanted to know if I could exist outside of
music. And it's been good. I can do other things!

I also wanted to know whether what I do is truly useful for society.
During the pandemic there was a lot of talk about essential workers,
like paramedics, truck drivers and farmers, but I've always had
doubts about my work. Everything we do in the world of culture is
considered a luxury. I work hard, it's a serious commitment, a lot of
effort, but I don't always know if it makes a difference. The same can
be said for art like the work by Eugènia Balcells. What's clear is that
music doesn't need me, and I don't need it. I look at it more like a
productive encounter. And if people like it or it gives them pleasure,
understanding, discovery or other kinds of experiences, then all the
better. If we look at it in light of Balcells' piece, we might say that
music is a part of the matter in the world. And, on my level, I give that
matter a voice.

As you learn to play an instrument, you improve your
technique and you're able to play more complicated
pieces, but it isn't all technique. There's something

deeper, isn't there? Maybe it has to do with that idea
of giving a voice...

If it were just a question of technique, no one would be interested in playing an instrument. There's a whole world behind every note, and everyone who enters that world will discover things – but we can never know which ones. I'm not a composer, my work is interpretation, and it's based on a score. From there, the sound gives you a form, a clue. It guides you, somehow. It's a very beautiful experience. I also teach a lot, and it's interesting to see how, starting from a common core element, everyone generates different possibilities. Music is an art of time, of moments. The same pieces – and this is fascinating – can give you very different things at different times. That's why, sometimes, I see myself more like a medium.

Clearly, what music conveys doesn't depend only
on technique. There are times when someone will
play a piece that may be technically complex, but
it doesn't speak to you at all. On the other hand,
there are people who may not be able to play such
complicated pieces, but they can play four very
simple notes that are brimming with emotion. To
me, it's all a mystery, but I've experienced it. I don't
know if it's related to that role as a medium.

It's a fascinating mystery that I experience every day. Music is an art of perception. What may seem cold to you could be very compelling for someone else. Music is like talking to someone through a medium. The performer is the medium for the music, which ties in with this idea of getting lost in the music. I love technique, and I can play very complicated pieces for hours and infuse them with a lot of diversity, a lot of energy, a lot of poetry, a lot of fantasy. But, in order to do that, you have to be willing to let the notes speak for themselves. That's why I was so struck by *Homage to the Elements*. In one of the videos I watched about it, I noticed a little boy staring at it intently. And that fascinated me because it demonstrates that the work speaks for itself. Matter can speak, just like sound and music can. Plus, the work incorporates a whole series of nuances of color that can also be applied to music.

Music, as you were saying, is a very effective vehicle
for communication. Sometimes you hear just a
few seconds, and it changes your mood because
you perceive an entire emotional world in it. The
philosopher Gottfried Leibniz said that music makes
the soul count, without realizing that it is counting.
In fact, music is an art of time that plays with our
expectations and can convey a torrent of emotion,
whereas other art forms may not be able to do that.
Is it a more direct art in that sense?

I'm actually a very visual person, rather than an auditory one. I'm more
sensitive to music than painting, but when I say that I'm a visual person,
I mean that when I'm listening to music, I imagine so much more than
when I'm looking at a work of art. That's why I prefer instrumental
music, because for me it already has a text. I think that music has
something ephemeral, transparent, invisible, that gives it that ubiquity,
that power to be everywhere, and also to envelop you and touch you.
Like the elements on the periodic table, music is vibration.

Why did you specialize in the Baroque?

That's a good question! I don't really know. Sometimes you start doing
something, and you're good at it, and life just keeps you on that path.
In any case, during that period, music had a very specific character; it
was very direct, a lived experience, and a very social element. Baroque
music has strong popular roots and it doesn't shy away from dance;
it's music to be shared and it has a lot of hope to it. That aspect was
lost a bit, especially in the 20th century. After two World Wars and a
whole lot of conflict, we lost faith in humanity. Baroque music is very
humanistic, based on a lot of beliefs and a lot of religion that I don't
share, but it also has a lot of faith in humanity. And that really speaks
to me. There is always an undercurrent of optimism, and humanity is
given the benefit of the doubt.

What is it like to play a piece of music that someone
wrote or imagined such a long time ago? How can it
be recreated? As I see it, in that sense the performer
is a lot like a scientist who interprets nature. In

truth, we can never really know what natural reality is like. Scientists have to settle for creating models to interpret it. The work of a performer also calls to mind an artist who explores a subject and captures it in a work of art, which, logically, can't represent that subject exactly or in its entirety.

I'm convinced that there is an important component of speculation or imagination in science. People who make discoveries or uncover new findings are actually engaging in a very artistic process. What I love about music is that all those aspects play a central role. The impression reigns above all, and I think that's why it offers so much freedom, so much room to play, and so much pleasure. Because no one's going to die if you make a mistake, you can try out lots of different things. Sometimes, for example, I imagine that I have a very strong connection with certain composers. It's like I know them, like they're whispering in my ear and I can contact them, or like they're speaking through me. It's true that there's also a very material, tactile, physical, terrestrial aspect. And there is an element of the unknown that will always be there. But there isn't any frustration because there's no one true outcome. And that pushes you to take more responsibility for your own evolution, to discover more things, to read more, to listen more, to open your eyes more, or even to close them.

You mentioned some similarities between music and Balcells' work. What did you see in *Homage to the Elements*?

First off, I think the piece is astounding. My first impression was just, simply, that I liked it. Because I like the colors, I like the rainbow, I like the idea of fragmentation, and I like the periodic table, which has a connection to my childhood because my mother was a chemistry teacher, and the elements had an outsize presence in our house. Balcells found this way of responding to the light that each element emits, and I see it as proof of the world, as something comforting that supports you and helps you understand your surroundings. I felt drawn in, as though the work was taking me into its arms and helping me. When someone points out something so simple, like she has, something so basic and universal that you had never thought about, it opens a door for you. Plus, I see a deep connection with synesthesia.

Once, when Franz Liszt was conducting an orchestra
in Weimar and heard a flaw in the music, he said
them, "Gentlemen, a little more blue please. This key
requires it!" He was a synesthete.

Exactly! There are people who see music in color. It's incredible.
There's an amazing case of a 20th-century French composer, Olivier
Messiaen, who is quite famous in the field of French classical music.
His mother published a series of texts, written when she was pregnant,
that describe a stained glass window, and then he created a piece of
music from it. It's a very profound piece. Messiaen worked extensively
with nature and with elements like birdsong. He composed immense
symphonies inspired by nature and he worked from the stained
glass windows of Notre-Dame de Paris to an incredible effect. And
I've always found that fascinating: the transmutation of a natural
phenomenon, like light or music, which are vibratory phenomena,
into other states. When I saw Balcells' work, it felt like someone was
corroborating that. To some extent, it seemed to me like an artistic-
scientific proof of things I had experienced.

Do you see colors in music?
I see them, but they're always transformed. Somehow, I internalize the
music with my body and then I transmit it. And that generates largely
tactile sensations, and even tastes. I get these pneumatic sensations
of density and palate. And I find that all the color in Balcells' work,
the changes in matter or in their voices, ties in with the atmospheric
pressure, as though some elements weighed more or less than others.
And then there are the black ones, which don't emit any visible light.

How do you interpret them?
I would say that there are some things artists can show you, and
some things they can't. In any case, it's as though she is acting as your
"eye for new things". And I think a musician can do the same thing,
because if I show you a score, maybe it doesn't say much to you, maybe
you can't hear it in your head. But I can. I have an internal way of
listening to it, and while that will never replace the sensory experience,
I can imagine what it would sound like. And I find there's a similarity
in that with what Balcells does because her work speaks much more

than the traditional periodic table, which also appeals to me in a physical sense. She turns it into something more alive, more attractive. And that opens up space for a discussion of matter that is less scientific and more personal. In general, I think we're too focused on ourselves. No matter how much we like nature or art or music, we're always talking about ourselves, how we see everything. And this piece offers the possibility of a more genuine sensory experience. It turns out that matter can talk to you. When Balcells says that light is the voice of matter, she says so because, all of a sudden, this matter that you may have never seen is speaking directly to you and, in a way, it gives you a break from yourself.

There is a French philosopher, a disciple of Martin Heidegger, Maurice Merleau-Ponty, who says that when you contemplate a sunset, you're always looking at the sun, but you never imagine that the sun is looking at you. And something similar happens with Balcells' work. And with music. Playing an instrument is more interesting when you let the violin, or the note itself, or even the light of the moment speak to you, rather than trying to produce everything yourself.

> That relationship with the object is very important in music, isn't it? Playing one violin isn't the same as playing another. The specific object you use, your violin, is fundamental because it has specific characteristics of smell, texture or weight that others don't have.

That's true, and it's incredibly beautiful. It's also true that you may be playing a different violin, but in the end, it will have your sound. And that's very interesting. The violin plays you, and you play it. We shouldn't look at the world around us just as objects that we use, or as products of our exploitation, because we're only here in passing but all those elements will persist.

> If light is the voice of matter, what is music the voice of?

That's an esoteric debate that has been going on for ages. The Greeks put music in the category of the arithmetic arts as opposed to the rhetorical arts. For them, its vibrations represented order in the world.

And music does, indeed, contain an order: if you stretch a string between two nails it makes one note; and if you pinch it in the middle, the note will be an octave higher. In a way, music is the voice of order or a certain tidiness in the world.

All this makes me think about what science does: it looks for patterns or regularities, because the world works a certain way, and there are certain things happen and others that don't. In the end, science uncovers this order or tries to recognize it.

I have a lot of respect for science, and I'm fascinated that there are people who dedicate their lives to investigating matter and finding order in the world. I think, though, that through music, through art in general, and, for example, through what Balcells does, we are more inhabited by beauty. I'm sure that scientists also get carried away by beauty, but maybe more intimately or more privately. For me, there's a difference between being a witness to the order in the world and seeking it out. I know that the world is ordered, I don't need to demonstrate it. Now, if I were a scientist, I can imagine I would find delight in the beauty of an equation or what Mendeleev did with the periodic table and the undiscovered elements.

Are you referring to how he predicted the existence of unknown elements in the empty spaces he left in the table?

Exactly. He knew that everything had to fit into it and that he couldn't demonstrate it. A lot of people said he was wrong, but in the end he turned out to be right. I think people like Mendeleev had intuition, they had genius, which is why I like science too, because it has that ambivalence between reason and intuition. People often assume that the most artistic people reject reason. That's not true! I work with physics. When I work with my students, I talk to them about vectors, directions, forces, gravity... And those are the fundamentals!

Homage to the Elements shows that each element emits its own unique colors of light, like a fingerprint. Could there be some kind of similar

mark in music? It's obvious that every instrument
has its own sound, but what about composers?
Are there features in their music that let you
recognize them?

Composers do leave their mark, actually. You can talk about their
particular way of walking with the notes, the harmonies and
rhythms, and, especially, that person's particular charm or DNA. But,
additionally, there's also a question of arrangement, tuning and scales.
Our tonal system hasn't always been the same; it has evolved, and
not precisely in favor of diversity. Temperament refers to the way of
arranging notes in an octave, and there are lots of different ways. In
Western music, almost all of them have disappeared and we play using
a system called equal temperament, but in other parts of the world
they use very different scales. In India, for example, if you sing a note
a little further back, they say it changes color, and then the meaning of
the note changes too. You might call it a nuance; we call it color. It's
very difficult to quantify, and there isn't a lot of vocabulary to describe
it. In France, for example, there is a current called spectral music. It
doesn't use the tonal system; instead, it makes more use of the concept
of color in music and the power of computers. Whereas in Balcells'
work light and matter enter into a dialogue, spectral music presents us
with a huge world in which music and light interpret one another.

Re-enchanting the World

Eugènia Balcells

The first time I visited Eugènia Balcells' studio, I was immediately sure of two things. One, that I would be a little wiser when I left, which turned out to be true – and has been the case every time I've been lucky enough to go back. And two, that I was dealing with a personality that could only be described as "volcanic". And I'm not referring to the common use of that adjective, which is often associated with an irascible character. When I use the term, I'm thinking of volcanoes as vehicles for, or originators of, creation. I'm thinking, for example, of the underwater volcanoes that formed the Azores eight million years ago. Because a volcano, ultimately, is essentially a channel for matter and energy that have a great transformative power. The volcano takes the rocks around it and uses the energy from the Earth's gravity to melt them, to process them, in a way, and then it dumps them into the outside world to create something where there was once nothing: an island that becomes a new landscape in the middle of the ocean, different from everything that was known before, and which will soon be colonized by all sorts of species that will evolve because of that volcano. And, like Darwin said in the last line of *On the Origin of Species*, they will give rise to "endless forms most beautiful and most wonderful". I don't think I'm exaggerating when I say that Eugènia Balcells has this ability to take materials, whether physical or cultural, and transform them with her overflowing energy into a new landscape – a landscape that is not just a work to admire, but a place we can visit, a geography that we can travel through not only to marvel at but also, and above all, to experience some kind of transformation. To appreciate this cosmogonic capacity, you only need to look at some of her work or talk to her for a little while. If you do, you'll find an open perspective that processes everything and which, combined with a special humility ("special" in the sense of species, not of unique) and a voracious appetite for any form of knowledge, has no choice but to surface as a creative force. The last time I spoke with her, before I wrote this, she was reading with fascination about the origins of life and microbiota – the 40 trillion cells with non-human DNA that live inside us and make us possible. I don't know what she must be reading now, but I do know that the power of an underwater volcano never rests, because creating worlds is in its nature.

How did you become interested in science?

There were two very important reasons. The first was that my grandfather was an architect and an inventor. He created lots of things. He was a cousin of Carles Buïgas, the architect who designed the Magic Fountain on Montjuïc, and we always had a very special relationship, a very strong connection. He was always inventing something. Once he made a pair of glasses with mirrors on the sides so you could see both what was in front of you and what was behind you. I remember sitting in the living room at home wearing those glasses and watching my family on one side and the buses, the taxis, and everything on the street, on the other. It created some strange circuits in my brain. Plus, I wanted to study philosophy and art, because those are the subjects that have always interested me the most, but my family wouldn't let me because they were afraid of my passion for film. In the end, however, after two years of private lessons in painting and various languages, I decided that it was better to go to university, which suited me well enough. I told my father I wanted to study to be a quantity surveyor (I come from a family of architects: my grandfather, my father, my uncle and both my brothers). Architecture school put me into contact with physics, chemistry and drawing. Years later, when I went to the United States, I had my degree validated, I did a Master's degree in art in Iowa, and I eventually studied philosophy there.

And the other reason?

The circular slide rule, an invention that my grandfather patented. It was a slide rule you could operate with one hand. Before that, architects would carry a linear slide rule in their pocket, but they needed two hands to use it. The circular model was much more precise and easier to use. When I was seventeen, I drew the mockup on a piece of plastic to serve as the model for production. It was my first paid job! Everyone laughed at us because the first calculating machines were already coming out, but I put all my energy into it. Years later, I realized that it was his legacy for me. All my work, in fact, stems from that circular slide rule, which uses a logarithm to turn multiplication into addition. Being able to look forward and backward at the same time, and the geometric structure of all the possible numbers that were contained in the circular slide rule – all that was incredibly meaningful for me.

It's funny how there are certain events or specific
objects that determine our lives.

The funny thing is, it all came full circle. Years later, I was in a very
serious accident. I had broken vertebrae and my back in several places,
my femur, my sternum, and I'd had an operation and I was in a cast,
and I had to spend a lot of time in bed. At the hospital they called
me Joan of Arc. While I was recovering, I couldn't even read because
I couldn't hold a book. All I could do was stare at the ceiling. Then
one day my doctor, Dr. Santos Palazzi, came to me and he said, "Joan
of Arc, I brought you something." He gave me a pair of glasses like
the ones my grandfather had made, but instead of having mirrors on
the sides, they were at the top. Using those glasses, I could look up
at the ceiling and read a book that was lying open on my stomach.
Coincidence? No, synchronicity. Something I had used when I was
little saved me when I was laid up in the hospital. At that point, the
process that had already begun in my brain took root permanently.
I've worked a lot with circles, moving forward and backward, and
with geometry and science. But all that was already there underneath,
waiting to come out.

How did it come out?

In 1979, at a bookstore on the Lower East Side in New York, I found
the book *The Tao of Physics* by Fritjof Capra. From that moment on,
I never saw reality in the same way again. Reality became a dance of
energy. Then I spent two years working from the roof of the building
where I lived, in Little Italy. I laid out a mattress, marked a spot, and
turned on my video camera. I explored every direction of the space,
during the day, at night, all hours... and I did a piece that was like an
electronic Stonehenge, called *From the Center*. I started it in 1979 and
finished it in 1982. In that piece, though, like in everything I've done,
in the end I didn't really invent anything. The only thing I've done has
been to process my experiences and what has presented itself to me.

How did you make the journey from that first piece
to paying homage to the chemical elements?

At one point, I was making a piece that is very special to me, *Color
Wheel*, which I devoted a few years to. I was always very taken with

Borges' image of the Aleph – in other words, a place which, somehow, contains everything. Obviously, I couldn't reproduce that, but I came up with an axis, a kind of *axis mundi*, four meters high and outfitted with translucent, rotating screens on which three projectors played a whole fabric containing strips of fragments from all kinds of films: industrial films, commercial films, Hollywood films, home movies, etc. Everything was classified by color in the piece because I was very interested in the power of color. And that also tied in with the fact that white light is a parade of seven horses, one of each color, moving together. I've always been interested in that organization. And *Color Wheel* was already organized in that way, which also corresponds with the arrangement of the chakras for Hindus, Buddhists, and much of the East. The seventh chakra, for example, which is the most open to another dimension, is purple. And notice that there is very little purple in nature, just a few flowers. It's the color of imagination and dreams, and, curiously, of the feminist movement. I always say that colors have a certain power. Red, for example, has the power of life.

Why is that?
Because all the red things on the planet are related to life: lips, sex, blood. Yellow, on the other hand, is the color of gold, the sun, and the food that comes from grain. Green is nature, leaves, trees, grass. Aqua is water. With blue you've got the sky, and dark blue is the night. Orange is the color of many fruits and of the skin of a significant proportion of humans.

And from that organization you moved on to
the spectra.
I've always been more interested in bridges than territories. And in my personal research, I've found a complex fabric of bridges that connect the vibration of an electromagnetic wave with Eastern philosophies and everything in the visible world. Over the years, I've come to discover that everything is one and that everything is connected. When I was finished with *Color Wheel*, I discovered the emission spectra and it blew my mind again. I can't believe the elements have a light signature! It's a dance of light! So it all aligns with what some Native Americans believe, that we came from the stars, and what some Eastern cultures say. Then, increasingly going out of my mind, I began to explore the

spectra of all the elements. I put them on the computer and made them dance, move away, come in close, so that they would blend together into a metaphor for the creation of everything. That was the beginning of *Frequencies*. And then, as an interpretive tool for the visitor, we did the periodic table using the spectrum for each element.

So *Homage to the Elements* emerged as a kind of key.
Yes, and people were absolutely fascinated by it, especially people from the University of Barcelona, in the Physics and Chemistry departments. That was the last thing I expected! Without thinking too much about it, just playing like a child, I had put together two icons that had been kept apart because of an excessive fragmentation in human knowledge. One from physics and one from chemistry. And I called it *Homage to the Elements* because I think if we don't give thanks for anything, we don't have space, we can't move. In our culture, we don't give thanks for anything. But if everything that exists is made up of just these few elements, they're the main players in all of this! And we hardly even know their names! So I wanted to honor them. That way we can get used to them and understand the outlines of their internal axes. I think we need to give thanks in that way. Unhurriedly, but we need to do it because it puts us in our place.

Gratitude seems like an important idea in your conception of the world and of life.
I started talking about it years ago in relation to the feminist movement, because it's something I have experienced very intensely. The work that women do caring for children, the elderly – caring for life, in short – isn't valued. But the world couldn't go on without it. The moment it's valued and given the weight it deserves, everything will change. In an instant. But no one talks about it. Honoring, which goes hand in hand with giving thanks, is a position we need to adopt in order to be able to re-enchant the world, which is one of the things that interests me the most. If I can contribute only a little to that, I'll be happy.

Throughout history, there have been many artists who have been interested in science. Salvador Dalí was one clear example.

Dalí was impressive, but we revolted against him because of politics, especially my generation. He had a very clear head and a keen interest in all of science. And the portrait he did of his wife, Gala, from particles, with that perspective, is one of the most exquisite things that has ever been done. He was a visionary. He also played the buffoon, of course, because an artist deals in things that are, in theory, worthless, but which are absolutely necessary, and that's a very awkward place to work from. He knew how to take advantage of that. I was never good at playing that game.

Lately, there are many artists who are looking to quantum physics for inspiration.
It's incredibly powerful. After relativity brought together time with space and space with matter, all that was missing was quantum physics to ultimately disrupt the material, ineffable, and immutable vision of a reality that we can control, that we can tame, and that we can use for our own purposes, which has left us in a very complicated situation. Instead of invoking, giving thanks, collaborating, co-participating, co-creating with a different attitude, of absolute respect, we decide to do something else. And we shouldn't be acting out of obligation or because of some rule, but because we're all here on the planet, because what's happening in the sea or on a beach is happening to me too. It's important to understand that we don't end at the limits of our skin and that we're creating reality together. The most important tool we have is the transmutation of our selves. Learning and changing until our last breath.

It's funny, because not understanding that we're all here happens to all living things. When they have abundant resources, they end up polluting and destroying their environment. The difference is that we're aware of it, and at the same time we aren't doing anything very relevant to fix it.
It's just that most solutions take an external approach, when they should be coming from the inside.

What do you mean?
I mean, we shouldn't be asking what we need to be doing with the

Earth and whether or not we should be dumping so much plastic into it; we need to understand it as if it were our own body. At that point it's non-negotiable; there's only one way to do things. And everyone can understand that. But we have a long way to go. I think humanity is in its infancy. We're starting to suspect some things, but we're still in the very early stages. And now we're faced with a challenge: either we do it right or reality will give us an ultimatum. We have to pay attention.

What is art for you?
There are lots of answers to that. The first is to ask yourself about art's role. From that standpoint, I'd define it as a platform, as a place – the only one we have on this planet without a predetermined purpose. And that makes it very fragile and dizzying, because it's surrounded by an abyss, but that also makes it unique, because it's the only place where nothing has been written before, where you can start off in any direction you want. Every time we've explored it, we've broken down barriers. Now, people even say that cooks are artists. Beyond that, I think that art is a very strange kind of condensation that reaches a very high level of energy within you, an energy that comes from a time period and that can be synthesized through a strange alchemy that transfers it as a legacy, and which, of course, celebrates beauty and knowledge.

So beauty is a fundamental and necessary result of art?
We have a very flat idea of beauty today, because we associate it with a kind of domesticated beauty, which is what society offers us. But that's just makeup. In truth, beauty is savage, wild, total. And we have to aspire to that. The Navajos understood this very well and they invoked beauty in an incredible way, with a poem that I really like. [She stands up.] With beauty before me, with beauty behind me, with beauty above me, with beauty below me, with beauty all around me. And that has to do with all the directions in space and with our physical body. It's an inhabitable beauty. And there is knowledge in that beauty, of course. It encompasses everything: the wonder and the difficulty or the tragedy of being human. That kind of beauty hasn't been hijacked by our culture, like the aesthetic pressure that is exerted on women, for example.

But there are artists who have rebelled
against beauty.

There has been a rebellion against beauty in recent years, but it's useless, because it's a rebellion against the beauty associated with an external patina, of making something pretty. But something pretty, if it isn't something strong, wise and which can open a window – even a very small one – onto a new unknown reality, it isn't anything. That's what art should do.

Jorge Wagensberg said that art, along with science,
and revelation or mysticism, were the three ways
humans have to access knowledge.

I'm a firm believer in all three. And art, at best, expands your vision a little. Goethe said that if we learn to look in the right way, any object creates a new artefact of perception in us. That means that, as observers alone, every object expands our ability to observe. In that sense, our vision is active. You create your vision, you shape it, it grows with you over the course of your life. And as a creator, the same thing is true. Every piece of work you do, every dialogue you begin, every little opening of a window lets you to dive a little deeper and broaden the next one. This mechanism, which is so difficult to define, is actually an attitude – an attitude of lifelong learning. As María Zambrano said when she returned from exile, "I'm glad to be here at the university with you, because I'm a student." The way our society works, it seems like there is a certain period in life set aside for learning, and once you've done that, you have to produce, reproduce, and then you die. But in truth, we have to keep contributing, until our dying breath, to co-creating this world and to making it more lucid and more peaceful, more loving, more beautiful.

There's a very short text of an interview with
the physicist Richard Feynman called "Ode to a
Flower", in which he criticizes people who say
that science, by trying to understand things, makes
them more boring and takes away their mystery.
That's something that has happened, historically.
In Newton's time, and until much later, there were

many poets who criticized him harshly for having
laid bare the origins of the rainbow.

Well I see it the other way around. I believe that knowledge broadens
us, because we'll never capture mystery entirely. When I read Capra,
who combined Feynman's diagrams with the dance of Shiva, with
all of Eastern philosophy and intuitive knowledge, as I see it, all that
broadened me. It didn't make me smaller.

Here, too, there's also the problem that, in many
fields, science isn't considered culture on the same
level as the humanities. To give a simplistic example
of what I mean, I would say that educated people
will all know who wrote *Don Quixote*, but they
probably aren't familiar with the structure of a
hydrogen atom.

I've always believed that divisions are disastrous, an impoverishment.
That's why I've spent so much time investigating bridges. Because
of them, we can inhabit knowledge and exchange it. I only know a
little bit about everything, but when a window opens up to me, it's a
cause for celebration. I remember how I felt, for example, when I first
heard the sounds of the planets. I had never imagined it was possible,
and suddenly I learned that NASA had recorded the electromagnetic
waves coming from each planet and translated them so they could be
perceived by human hearing. That's when I did *Universe*, to put all that
in a public place so people could sit down and listen to it. Ultimately,
it's knowledge that doesn't make Venus seem smaller to me; on the
contrary, it makes it bigger, more interesting.

As you've mentioned several times, throughout
your career you've been greatly influenced by
cultures from other parts of the world, especially
Eastern cultures.

Absolutely. They've influenced me because they are very old and
rely on very broad concepts, which are highly relative and not very
dogmatic. For example, Hinduism or Eastern philosophies encompass
everything. Either everything is sacred or there can be nothing
sacred. And I understand that. There are many different levels of

consciousness. We evolve and we become a little more enlightened and a little less material. Matter and light are the two extremes. Matter is slower and heavier, whereas light is very fast. And here's an interesting exercise: when we look at the sun, we're seeing the light it emitted eight minutes ago. So everything we're looking at, we're seeing in the past. Even if it's just a millionth of a second, I'm seeing your past. And understanding those questions expands our minds a lot, and that has advantages. Because then we can make the decisions that we need to make – both the ones that affect us and the ones that affect the planet – from a wiser, more complex and more possible place. Otherwise, we'll always be stuck on the same plane, which is the case with so many entrenched conflicts, like the one between Israel and Palestine. We need to let things go right away so we can be alive in the present. And we need to do it both individually and collectively.

But that's very hard to do, isn't it?
It's work that comes with the appetite for life, putting things in order every day, letting go of things, not sitting in judgment. Because when you refuse to budge, even if you're defending a position you believe is right, you're doing the same thing as the other side. And you're just digging a bigger and bigger hole.

Plus, in the Judeo-Christian cultural tradition there is always judgment.
There is judgment and there is grievance. The idea of "what am I owed?" But, of course, a debt can never be resolved. And yet, all the possibilities for the future are always open. The problem is that a perspective that focuses exclusively on the idea of debt restricts our thinking, and it means that when the debt is finally paid, you can still be a thousand times poorer.

What you said before about art being a kind of condensation of the energy of a period, I think it has a lot in common with what science does – for example, in the case of synthesizing all the gravitational phenomena into one equation.
There are a lot of parallelisms, but art incorporates metaphor, the tactile

element, sensuality; it allows a lot of space for intuition. In other words, art falls somewhere between mysticism and science. Because there has to be a certain openness to things that are unknown to you but that you let in. Picasso said that when he was working he was always willing to be a medium, a conduit. Often I'm aware that the best part of the pieces I've done reaches far beyond me, that I didn't put it there in a controlled way, but rather it was something that wanted to be there, and it came to me because I was receptive to it, in a sense. And then, as a spectator, when I see that, I'm always taken aback.

Where does it come from?
It's a gift, a blessing, a revelation. I don't know how else to put it. It's what Wagensberg needed to put in that other category. There are certain artists who have more of it, and others who simply don't. Because it demands humility, acceptance, being very open and not having too much of an ego. If there's an open channel, it can happen. And then, of course, it is put through the artist's filters, and they decide what to do and how. But I've always said that the best things I've done aren't really mine. That's very clear to me. And that gives me a lot of peace of mind, because it takes the ego out of the equation. Although saying that isn't very well received. At best, art is a window onto a certain revelation.

There is a kind of art that is difficult to understand and interpret.
Of course, a trained eye can see a lot more in it. That's why I've always tried to use very universal codes. But it's true that the codes in the art world have tended toward becoming more closed off. If you don't know who Marcel Duchamp is, you can't understand a piece that consists of three suitcases in an empty space. And using restricted codes has made the art world very narrow; it has done a lot of harm. Those hermetic codes create a world for art instead of art for the world. Initially, art, like knowledge, belonged to the world, but lately, a world for art has been cemented. Those codes have been trafficked along with a series of ideas about museums, galleries, contacts, money and power, and the only thing a lot of young people are getting in schools now are the codes, instead of the art.

Codes, however, are part of the language we need to
understand any intellectual development.
In the case of art, they are a language, but one that is closed off.
A language that ends up being more like a masturbation because
everything refers back to those codes. And that gets certain experts
talking about it, but when it comes to outsiders – cultured, educated,
intelligent, well-traveled viewers – at the end of the day, they aren't
interested in art. In recent years, and I suppose this has worked against
me when it comes to really being a part of the art world, I have
wanted the codes that I use to be as broad as possible, so that there is
a possibility for the works, in one way or another, to be a part of life.
That's also why I work with science so much, because it has more
universal codes.

In the end, all of this leads to a kind of banalization
of art, where many people don't value contemporary
art because they think they won't understand it.
It's like a joke. And no one makes sure that it opens a small window
onto understanding reality – even if it's very narrow, but it has to open
something. Art can't help but create those openings. The first one to
do it was Duchamp when he took a urinal and put it in a museum.
That action opens a window because it's saying a series of things about
art. The second after he did it, it almost wasn't worth it anymore, and
after it has been done 200 times, it has no interest whatsoever. There is
a bit of a farce to it. You have to be very careful.

Earlier you mentioned that you were reluctant to
say that your works are not entirely yours, but that
there's an aspect of revelation. Why is that?
Because people tend to lash out against mystery, against anything that
isn't tangible, measurable and weighable. Religions don't help in the
matter, because they've become structures of power and norms, but I
advocate the human spirit in all its forms. And mystery. We aren't just
mind and body; we're also emotions and spirit. And that's where life,
health, creation have to come from. That's why we need to take care
of our spirit and our emotions in the same way that we attend to our
bodies every day. For example, one pattern that we've got completely

wrong is the idea of how we live with respect to our age. We're born, we grow, we hit our high point at forty and then the decline begins. Carrying that around inside is worse than an atomic bomb. We need to erase that from all our cells and assume a curve that just keeps rising. With that curve, we can all enter into a dialogue; with the other one, someone on the way up has nothing to say to someone on the way down, who is in decline and has no space to evolve or to contribute anything. It's crazy! Accepting this new curve would change everything, especially for women, who are heavily dependent on image. Instead of slipping toward an end of decay, loneliness and misery, we could reach a moment of plenitude and knowledge in which we would give thanks and then, in an act of complete love, we would stop breathing and let go.

Is that possible?

There are masters of Hinduism in Bali who do it. We might not be able to do that exactly, but we could get close.

When you talk about mystery, what do you mean exactly?

Everything that I can't know. But even if I can't know it, I don't deride it or denigrate it, just the opposite. I honor mystery, I coexist with it, I embrace it. And if it wants to reveal itself to me a little, fantastic. Life is a mystery. Isn't it? The fact that you and I are here talking now is so improbable. Everything that had to happen to make it possible is incredible. But we're here talking, and it seems like we're communicating. There's no greater mystery than that, that we're here, that we have this strange form, that we move, that it seems like you're listening to me, and that we're creating a rhythm, a little dance of questions and answers. What more thrilling mystery is there? And honoring that mystery is also important. We may not understand it much, but if it's a good moment, a lucid moment, let's be thankful for it.

Science has advanced, and our understanding of the world has changed. Maybe that's why this kind of mystery has lost importance in Western culture?

The problem is that we've destroyed it. For example, we've annihilated the Native Americans. We destroyed so much ancient wisdom, but luckily it can still be recovered. In that sense, I had a moment of clarity a few years ago. I was traveling through the Albufera coastal lagoon in València, and I saw that they were planting rice. And I said to myself, "Now I get it!" My assistant asked me what I meant, and I said, "Rice is planted with rice," which became the title of a film I made. He laughed. "That's ridiculous," he said. Precisely because it seems ridiculous, it's so completely profound. If you don't have rice, you'll never get rice. If you don't have love, you'll never get love. Rice is planted with rice. Creativity is planted with creativity. To get some, you need to already have a little, or someone has to give you some, and that means you have to be very humble. The ancient alchemists wanted to make gold from other metals. But if you don't have gold, you can't make gold.

Plus, there's a limited amount of gold and other elements on the Earth; we can't make them.

Exactly, it took cosmic fires to make them. If we run out of iron, there isn't any more. If we run out of gold, there isn't any more. We have a limited quantity of those elements. In a way, it's our legacy. And when you realize that, it makes more sense to celebrate all the possible combinations, from a single atom to the brain of a complex animal, to celebrate the whole journey that evolution has taken over millions of years. At the same time, it cuts us down to size, and it gives us a point of reference and a reason to be grateful and to want to contribute the best of ourselves toward rediscovering the enchantment of life.

How did we lose sight of that enchantment?

Through pain. We've killed too much. There has been too much violence. If you're looking for a film on TV that isn't about pain, war or desolation, you won't find anything, because we've become addicted to those vibrations. Things that are too lightweight can't hold our attention. Even making love softly bores us. We're so desensitized that we need the greatest wonder there is to be harsher and more extreme. The past century, with its World Wars and the killing of so many innocent people, left a trail of bitterness, negativity and violence. We're suffering from a collective disease, and the people in power are the sickest.

Can we do something about it?
We need to take as much power as we can and use all the tools we have – art, science, poetry – to encourage one another. In the end, we have to dance this dance together. Together we need to co-create reality to make it into an enchanted place. And today more than ever, because we're in more danger.

A Factory
for Freedom

Federico Mayor Zaragoza

I feel like, sooner or later, everyone who works in science ends up thinking about freedom. And when I say thinking, I mean exploring the foundations of what we call free will. If, as people were imagining in the 19th century, there are natural laws by which the evolution of any system can be predicted by looking at the initial conditions, then human freedom doesn't exist. Although we are a system made up of trillions of particles and it is impossible for us to make concrete calculations, according to that perspective, everything we do is determined by those laws. Our desires, our struggles, and our passions are inconsequential. The laws of nature will win out. Or rather, our desires, struggles, and passions will also be a consequence of those laws. Like a billiard ball that misses the hole by a millimeter. At the beginning of the 20th century, however, the study of matter on the atomic level led to the development of a new theory, quantum physics, which runs contrary to that deterministic vision of the world and reserves a space for freedom. In this new paradigm, the properties of physical objects aren't determined until they're measured, and when that happens, it can result in different values – not just one. Some people have seen this as a resurrection of the importance of desires, struggles and passions. One of them is Federico Mayor Zaragoza, who has thought a lot about these questions, and, what's more, has done so not only from within scientific paradigms, but also from the perspective of political action and reflection. Throughout his long career, he has come to the conclusion that freedom exists, and that art is one of its factories. The same could be said for science, I would add. Because both art and science construct a vision of the world that is new and more profound. And as Federico Mayor Zaragoza knows, it is there, in our vision, where true freedom lies.

When you first saw *Homage to the Elements*, what were your thoughts?

I thought to myself that light has a huge influence because, in the beginning, there was only light and, maybe at the end of everything, there will be only light again. On the other hand, I think it's incredibly interesting that the elements have a connection to light like we have with our DNA, which identifies us individually – the fact that each element has this fingerprint, this light-based symbol – because, after all, our goal as scientists is contemplation. And contemplation requires a series of guidelines like the ones we find in this piece. Guidelines that lead you to think in a different way. When I was working in the biochemistry lab at Oxford University in 1966, I had a memorable conversation with the Nobel laureate Hans Krebs. I told him that I was getting a lot of data, and he immediately warned me: "Be careful. Having a lot of data is dangerous because there may come a time when you start to think that research is just collecting data." And then he said something that ties in very well with Balcells' work: "Research is looking at the same things other people are looking at and thinking something no one else has thought."

What new thoughts do you see in Balcells' work?

She offers a masterful interpretation of the fact that matter can be expressed in terms of light, and she does it with a show of great creativity, which is the most important and distinctive capability of the human species. Creativity can open up new pathways to imagine different futures. We often think that the most important thing is the device or the tool, whereas the important thing is the idea, the reflection, the anticipation, the imagination, which is just overflowing in the case of Balcells. But, as I said, of all these qualities (thinking, imagining, innovating, etc.) which are specific to humans, creativity is the most outstanding. If I were asked how to define Balcells' work, I would say that in all her descriptions of the universe, matter or the elements, from the largest to the smallest, what always stands out is her creativity. From a biological standpoint, creating is superfluous. The fact that every human being is capable of creating is a justification for human life in and of itself.

> People tend to see science as an activity that is cold,
> rational, calculating and inflexible, but creativity also
> plays a really important role in it.

Thinking what no one else has thought, which is the foundation of research, is a creative act. On November 19, 2002, I wrote a poem about creativity which, like Balcells does with each element, attempts to translate the whole universe into a thought, a way of thinking. It's very short, and it goes like this:

> *The universe fills the mind with unanswered questions*
> *perfect, immense, infinite*
> *so that everyone can believe*
> *so that no one is forced to*
> *so that they can densely live every moment of the great mystery.*

And in the face of that mystery, we can't limit ourselves to just looking at it. With her work, Balcells is telling us that when we look at something, immediately we should look at what may be behind it, and why it exists in that way as opposed to any other. How might we hold onto the pathways of those thoughts and reasonings? I think *Homage to the Elements* holds that secret.

> Does that capacity for synthesis, formalizing
> all those thoughts into something that can be
> communicated, require the language of art?

Yes. That's why I always say that Balcells' work doesn't just incorporate science but also philosophy, sociology and, above all, creativity. In the end, everything is based on the grand solution, which is for all human beings to design their own lives, for all human beings to realize that not only can they reflect but also create, and that they can each invent their own future. We have to try to put into words what no one has said yet and, above all, never act in keeping with anyone else's dictates. That's the most important idea I find in a piece like this one. Moving from matter to thought, from thought to creativity, and from creativity to freedom.

> Does that mean that art has a moral dimension?

Art is what it creates. And I see a very clear relationship, a kind of triangle, between philosophy, art and freedom. Freedom is a supreme gift. When we lose that by following the thoughts or dictates of others, we succumb, and instead of being free, we become submissive. And Balcells extols freedom because she can understand that both infinitely large and infinitely small things are moved by light, and they can be represented in an extensive work of art of incredible quality. Ultimately, the conclusion is that our hope lies in every single, free human being who is capable of creating.

Is this kind of thinking, elicited by works like Balcells', present enough in the political sphere?

No. That's why I think it's so important for Balcells' work to be disseminated more widely. We're tired of just looking on, of being simple spectators, without engaging in this kind of reflection about the work. Today, social media has given us a certain freedom. For the first time, we can all express ourselves, and that's wonderful because it wasn't possible until very recently. Now we have the floor, we have a voice. But, of course, we have to exercise that faculty, because if we remain mere spectators, we aren't playing an active role. The great thing about *Homage to the Elements* is that it shows us how, through art, through the image, through each element's light signature, there is an illumination of other aspects of life. We can't just go through life saying what we're told to say and doing what we're told to do. Freedom is precisely the consequence of knowing that any one of us can make a different design for tomorrow.

How does this idea of freedom relate to morality? In other words, does freedom need to be limited in order for life in society to be possible?

Freedom is moral in its essence. When someone oversteps the limits of the supreme gift of being free, they are no longer exercising their freedom adequately. In that sense, Miguel de Unamuno, one of the authors who reflected the most on the subject, said that "freedom sits on the exact edge of uncertainties and certainties." And if that freedom is used to influence others so they'll act a certain way or do things they don't want to do, it's a misuse of freedom. And morality, belief, and dogma tend to go hand in hand with the illicit uses of that great power every human being has to be free and creative. Right now we find ourselves at a juncture where we need to realize that we have to change – that we have to invent the future. It's unconscionable that there are people starving every day while millionaires are going on space flights that cost a fortune and are absurd from an environmental point of view. How is that even imaginable? Not by making use of freedom, that's for sure. And on the other hand, if we apply moral rules that someone else has made up for others to follow, we aren't making use of freedom either. And in that little poem, I wrote that everyone should be able to believe, but no one should be forced to. That's freedom.

Are we freer now than in other periods in history, or are we too dominated by technology?

While it's true that we're letting ourselves be controlled by technology, right now there can be no doubt that, for the first time in history, human beings are able to gain access to the full exercise of our unique faculties, including freedom. When people criticize the work we did during the Transition, I like to remind them that at that time, 45 years ago, only 10% of Spanish citizens were involved in what we might call the Transition or political activity. The other 90% were born, lived and died within the space of 40 square kilometers and knew nothing about what was going on beyond that. I remember a similar situation during my youth in Barcelona, which wasn't a tiny village: we didn't know anything about the rest of the world. To begin with, women were invisible. There weren't very many cases like Balcells because women were completely marginalized. And in politics it was all men. But I kept saying: listen, this isn't right, there are just a small group of us making decisions. And thanks in a large part to digital technologies, that has changed. For a little over 20 years now, most human beings have known what's happening beyond the confines of their own lives. And above all, women have started down the road to full equality with men in every area. That's essential, because that's the pillar of all human rights: the idea that all human beings are equal in dignity. And today we have to acknowledge that women and men are equal, regardless of their sexual orientation, beliefs, ideology or ethnicity. Everyone has the same dignity. And as a result, we can be free.

How does art contribute to that freedom?

Art is a source of examples of how to look at things in a different way, through a different lens, like Balcells does, and to give that fundamental importance to creativity and self-esteem – in other words, to act according to one's own thoughts and beliefs and not according to what other people say or think. That's essential, and now it's possible. That wasn't the case 30 years ago, but it is now. It's possible for each of us to make our voices heard and to make use of our freedom. We weren't making use of our freedom. And so, of course, we were silent, fearful, obedient, submissive. Imagine to what extent that's true, that the first paragraph of the preamble to the Universal Declaration of Human Rights states that the full exercise of those rights is meant to free humanity from fear. Everyone used to be scared, and, what's more, there were just a few men telling us all what to do. And everyone else was expected to obey, especially women, and that was that. But now, for the first time, we don't have any more reasons to be afraid. Moreover, we always have an avenue that a lot of people would like us to forget: the declaration also says that if someone hinders the exercise of those human rights, rebellion can be justified. It tells us that we aren't slaves, that we shouldn't be submissive or obedient. If we can't exercise our human rights, we should rebel. That doesn't mean the rebellion has to be violent. Rebellion isn't the same as revolution. But we can defend ourselves; we can say no. That's why I think Balcells' work is so important, and I think we should thank her for helping us to look at nature in a different way, to reflect on the universe in a different way, and think about the infinitely big and the infinitely small in a different way. I think it's essential for that vision, which confers creative capacity and freedom, to be recognized.

> Despite improvements, we don't have complete freedom. There are intense inequalities between countries and, in the case of women, there are still gender biases, even in developed countries. Most of the important positions in business and in the political sphere, for example, are still held by men.

True, but we're past the point of no return. Forty years ago there wasn't a single woman in the Spanish government; now there are more women than men. It's true that we still have a long way to go, and there will always be something missing on the road to full freedom – especially since there are lots of people who are interested in obstructing it. Those are the same people who defend religion and the kinds of thinking that foster blind obedience. There's a problem with supremacism in Europe today, and supremacism is what led to the Second World War. When Adolf Hitler wrote in *Mein Kampf* in 1933 that the Aryan race was incompatible with the Jewish race, he was laying the groundwork for what happened later. Then Benito Mussolini said that the Romans shouldn't be inferior to the Aryans. And then Hiro Hito emerged with a similar discourse. We need to be on the alert.

What can we do?
The key is in our everyday behavior. And that's why, looking at a piece like Balcells', the goal shouldn't just be to extract new and beautiful concepts from it, but to derive resolutions related to our everyday lives, to how we lay out our daily routines and decide on our behavior – how we live our lives, more generally speaking. And that's the most important part of culture; that's what we should be advocating now. We can't accept leaders like Donald Trump in the United States or Jair Bolsonaro in Brazil, who are opposed to freedom and who would impose hegemonic principles or rules of conduct based on religion, which instead of letting us soar, clip our wings. The world can't go down the road of supremacism, dogmatism and fanaticism again. Not again. At the same time, there has to be equality in dignity, freedom and creativity. The Nobel Peace Prize winner Rigoberta Menchú, who is a Mayan from Guatemala, once gave a speech that astounded me. As she was finishing, she said that every day, when we wake up, we should dedicate a few minutes to thinking about what we're going to do that day, laying out a roadmap. If we can't do that, she said, then we should apologize like in the holy book of the Maya: "Forgive me, dawn, for not receiving you as you deserve." I hope that, after seeing Balcells' work, there will be many more people who will be capable of receiving the dawn as it deserves.

Another Narrative Is Possible

Simon McBurney

Simon McBurney looked familiar to me. As we talked, I tried to remember where I had seen him before. I finally figured it out. He was Attlee! The high-ranking operative from MI6, the British Secret Service, who engages in all kinds of obscure exploits and mysterious maneuvers in the film *Mission: Impossible – Rogue Nation*. Spycraft, in short. I could hardly have imagined that the person I had first seen in one of the action film viewings that we occasionally indulge in at home was also the author of the play *A Disappearing Number*, about the relationship between the English mathematician G. H. Hardy and the self-taught Indian mathematician Srinivasa Ramanujan, or the suggestive piece *The Encounter*, which, through three-dimensional sound and a well-honed script, transports theatre-goers into the amazon rainforest, as they ask themselves all sorts of questions about how we live in this world. I have to confess that I'm a bit embarrassed I had only heard of him, at least for a time, from a spy movie. Simon McBurney is much more than a spy. Or maybe he's a spy of human nature. But, above all, he is someone who understands creation in a very broad way, and who reads and thinks a lot about his art. A kind of art that expresses things that can't be put into words. An art that explores what exists beyond words but which, in some way, we all carry within us.

What were your first impressions when you saw
Homage to the Elements?

It's a fascinating piece. Absolutely fascinating. We tend to associate
the periodic table with the names of the elements that make up our
universe and the symbols we give them. But seeing their colors gives
them an identity that falls outside our language. And that's fascinating
because you might say that the elements almost come to life as
characters. I'm thinking of carbon, which is very interesting because it
doesn't have very many lines. Some elements have a predominance of
blues, reds or yellows, but carbon is very balanced and it's very simple.

In your play *The Encounter*, you portray a tribe from
the Amazon, whom you personally interacted with
and who, in a way, give a different formulation of the
idea we see in *Homage to the Elements* – that we're all
made up of the same elements. Their consciousness
is different from ours in Western societies: they
believe they are made up of things that are physically
outside their bodies, like plants and trees.

To answer that question I first have to offer a few preliminaries on the
idea of consciousness.

Please.

Before *The Encounter* I had worked on a piece that explored the
relationship between memory and identity. Who am I? Am I the sum
total of the memories I have in my head? And do those memories
create what we call consciousness? Is consciousness possible without
memory? Some people say that we can be alive and conscious even
if we're suffering from dementia or another disease that erases the
memories from our brains like Alzheimer's. But that's probably
because there is still a remnant of memory. So memory is absolutely
essential for consciousness. However, if we think of consciousness
in a simplified way, as a function of the brain, we fall into the trap
of separation. We start separating the function of the brain from
everything else. The brain is naturally connected to the body through
the nervous system, which feeds it information that it then uses to
create memories and to begin building a consciousness. If, as a child, I

touch something very hot, I store that event as an unpleasant memory, I learn from it, and I begin to build synaptic patterns, which become more and more complex until I have a consciousness of who I am. This idea highlights the fact that consciousness is formed through bodily processes, meaning that the body and the mind are inseparable. And our bodies are in constant contact with the outside world. I touch the outside world, which stimulates my nervous system with an impulse that travels to my brain and adds another tiny piece of the many, many pieces that make up my consciousness. Light enters the eye in the same way that a hand touches a hot plate, and registering that plate tells us that what we see also constructs our consciousness. Light, therefore, helps construct our consciousness. In the same way that the things we smell do, as well, because smells come from the outside world. But there is a fascinating fact about this: we can hear lots of different tones, our eyes can perceive an extraordinary number of colors, but our sense of smell perceives not just millions but billions of different smells. We have a very developed sense of smell, much more than many other animals. We tend to forget about it, but its complexity is extraordinarily important when it comes to defining who we are through our consciousness – perhaps more than we recognize because we are often unaware of its true scope. And then we have our ears, which, unlike our eyes and the light that enters our consciousness, can never be shut off. Our ears are open all the time, and the brain has an enormous job, which is to inhibit and select what we hear. We've all had the experience of being in a crowded room and concentrating on a conversation with someone who may be a few feet away, and yet we're still able to filter out the noise around us and hear what they're saying. In a similar way, when we're at home in cities like Barcelona or London, we inhibit the constant traffic noise. During the lockdown, the silence and the birds singing came as a big surprise. That strange silence alerted our brains to something new.

What you're talking about happens in London and Barcelona and in the Amazon rainforest as well.
Right. But the people I met in the Amazon, the Xingu, whom I consider friends, had a very curious answer to the question about where consciousness resides. I often bring up this question as an

exercise when I work with actors and other people. The most common response is that consciousness resides behind our eyes. But the answer from many people in the Amazon was to point toward the jungle. Initially, I thought there had been a mistranslation, so I asked the question again in a different way, with a different translation. Finally, though, I realized that I was the one who hadn't understood. For them, there's no boundary, no barrier, no limit or separation between what happens outside and what happens inside – or what we consider to be inside our brains, in those 1,200 grams of electrified pâté.

It's a recognition of the fact that, when we're little and we touch something, we feel that it's hot and we start to register that, what is inside and what is outside are part of the same whole. For the Xingu, the complexity of our interior is reflected in the complexity of what is outside us. So the harmony of what they see can be reflected in the harmony of what they feel. Likewise, if the brain is unbalanced in some way, it can also be reflected on the outside. And here they offer an important reflection on our current world, where they see Western culture destroying its environment, destroying huge areas of rainforest, ripping things out of the ground and dumping waste into the water... It all leads to an external devastation that could be the reflection of an interior devastation in our minds and our consciousness. I often think that that ugliness, that disconnect, implies that we are also disjointed ourselves, and that we have scars that demand some kind of treatment.

> This idea of unity, which without originating in science coincides with it in some respects, must have moral and social implications because it is more internalized, as opposed to being a purely intellectual construction.

One of the most interesting consequences of this deep understanding of our continuity with the outside world, the fact that there is nothing between our skin and what is outside, is that there's no gap between Toni Pou and Simon McBurney. Both are continually flowing from one to the other, such that social decisions can never be made on an individual scale, but only on a collective scale. The Xingu think as a community. They believe they are a collective consciousness, and they are deeply aware that their decisions should be made as a group. As a result, their sense of community is like that of an organism as a whole – like an element that is made up of all these different lights that form a whole; they aren't separate pieces of light but a unit. Plus, they have a deep understanding of the fact that as you grow older, you have more memories, which, in turn, is important to everyone in the community. The accumulated memory of the elderly is absolutely critical. And the strength of young people is equally essential. That's why one of the terrible consequences of COVID in the community is that it affects the elderly. As a result, the Xingu have begun to lose what they consider to be a vital component of their collective consciousness: the memories of those who have lived longer than everyone else. My friend John Berger, who has also been very important to me as a writer, wrote a very lovely phrase about that: "Nothing can take the past away: the past grows gradually around one, like a placenta for dying." As if we had this protective wrapping that lets us to move on to the next stage of development. And it was very profound to feel that in the Amazon. When I talk about it, I wonder if it wouldn't be better to use the word *learn* as opposed to *feel*, because one of the things my experiences there helped me to understand is the extent to which I am a prisoner of my culture and my sense of consciousness. To really understand and learn something like that, you have to have the ability to change and feel differently. And that's an enormous challenge.

This way of understanding the world and themselves also ties in with the idea that death is another phase in a cycle that is larger than human life, which is something we can also take from *Homage to the Elements*. In Western cultures, on the other hand, we are very anchored to our own lives and approach death from a place of fear and sadness.

We live on a human scale. The human body sees the things that relate to that scale. As humans, we can't see infrared light or ultraviolet light, but we've found tools that make it possible. And yet, we can see rainbows, we can see how light breaks up into its constituent colors, we can perceive light even though we can't perceive all of its qualities. We can experience the speed of light in a valley when we see someone shouting and we perceive the image faster than the sound. We see an ax hitting a log or one stone crashing into another before we hear the blow. At night we look up at the stars, and we begin to wonder what's out there, what that world is like. And I'm think of us as human beings before we had access to so much data and so many tools that enable us to see farther. Around us, there were the animals, the trees... In other words, if we situate ourselves within the world that is actually around us, death is everywhere. Life is everywhere. And they are both absolutely and constantly interconnected. On the other hand, if we are separated from our surroundings, or if we don't dedicate any time to looking at them or being present in them, death begins to manifest itself, apparently, only in terms of ourselves or the people around us. And on that simple level, we see it only as something that affects us. But if you are constantly on the circumference of the totality of life and death, you understand that you are part of an unbreakable continuity and you have a profound understanding of how the death of one thing contributes to the life of another. The moment you fully understand that the dead, death, and our dead are all inseparable from the living. Death itself is part of life because it is our ultimate future, and the dead and the living are bound together because they are part of everything that we have experienced. John Berger said that the dead include the living in their collective, that the dead surround the living and that the living are the core or the heart of the dead. In this core are the dimensions of time and space, but what surrounds the core is timelessness. And the dead exist in this place we might call timelessness. In that sense, we coexist constantly with the dead. The dead are present for us and, and they shape, if you will, our imagination. What Berger says is that there is a constant connection between the living and the dead, and that it has always been there. But he also says that there is a unique form of egotism that has broken this interdependence. A uniquely modern form of egotism. A kind of narcissism, perhaps, with disastrous results for the living. Because now the living think of the dead as eliminated, as no longer present.

> How is this more circular conception of existence
> ritualized in the Amazon?

Precisely one of the tribe's fears regarding COVID-19 was uncertainty about what would happen to their dead. Because they have an unbreakable bond with the dead, it is important for them to be involved in the process of death. And that bond is broken if the bodies are simply cremated somewhere far away and can't take their place in the community. They usually bury their dead and, after a full cycle of seasons, they dig up the bones. Then they have a very special ceremony, which gathers all the nearby communities together. The bones of the dead are buried in the center of the village, A tree is cut down, and painted, and it is set it up like a totem, as a symbol of that person. The family mourns all night, accompanied by others. There is continuous dancing and singing, recreating the legend of how fire was given to humankind. The story is told and represented by running through the forest carrying fire. When the sun comes up, the mourning is over. The year of living with that person's dead reaches a climax, and their spirit is released to be a part of the periodic table, the world, and the universe. Then they knock down that beautiful totem and toss it into the river to float away. Any feelings of sadness or helplessness are quelled because the dead are part of the continuity that surrounds them. The mourning focuses on the rupture and the wound, and their presence is celebrated. And COVID-19, a Western disease – for which there are historical precedents in their experience – has been terrible. It has caused a rift in the sense of continuity and connection with the world around them. My Xingu friend Takuma Kuikuro, who is a filmmaker, says that their consciousness is double now: there is one in the village, and another called white consciousness, which is the consciousness of the Western world.

> In Western cultures we separate everything – inside and out, before and after, life and death – but as you were saying, as the Xingu see it, things aren't separate. Do you see any connection between this fluidity and the universal quality that is transmitted by *Homage to the Elements?*

I think a lot about how, in the Western world, we've come to believe in a narrative that excludes many others. The great thing about the elements from the periodic table is that, by themselves, they don't do anything. They only do something when they're combined. Our thinking today is part of a narrative that might be called the narrative of *logos* or logical thinking. It has grown thanks to the impressive evolution of science, which, with its observations and empirical evidence, has revealed so much about the mechanisms that create life and describe the vastness of the universe. And, above all, since the 19th century it has shown us the immensity of the past. Two hundred years ago, no one imagined our past was so huge. That's why the discovery of dinosaur bones was such a big mystery. The discovery of the immensity of the past, which is very recent, has taken some of the weight off our being. And the discovery of evolution reveals the construction of a world that leads to the appearance of Toni Pou's brain and the construction of the cells that make up his mustache and his beard, and it is all such an extraordinary narrative that reveals so many things... And like all scientific narratives, as it opens further and further, and we perceive smaller and smaller fragments of the universe and we understand larger elements, more and more mysteries emerge. Have there been universes before ours? Are there parallel universes? If there is only one, does it have an end and a beginning? And if that's the case, what was there before the beginning? In that sense, I think *Homage to the Elements* and *Frequencies* both remind us of other narratives. Carbon, for example, is the prerequisite for life, but it isn't enough on its own. It has to relate with other elements. Even if we separate the periodic table into individual elements, it's clear that they aren't essentially individual. And that's what the piece reveals, which I find so beautiful... And that's where I see the relationship with universality. To understand the elements, we've used this logical process of reduction, deduction and separation. But then, we find that each element has to be combined with other elements to create life.

Do those excluded narratives come from the world of art, mythology, religion?
When it comes to understanding something, there is always more than one narrative. And alongside the kind of human understanding that functions according to the narrative of *logos*, there has always been the narrative of myth, which is about finding a way to describe the things that exist beyond words. If we look at it from the perspective of science, we know that there are things that escape our current understanding, and we imagine that we will eventually come to understand them.

But in any case, I think we are at a turning point, and we are beginning to see the limitations of our logical thought process. We see it in how we treat everything around us, in the enormous rise in mental illness, in our growing inhumanity when dealing with other animals or plants, and, more generally, the planet as a whole. I would call our behavior unscientific, because using an extremely toxic pesticide isn't listening to science – it's quite the opposite, because it implies ignoring the consequences that come with it.

> What role does the narrative of myth play in all of this?

We all need to understand our own mortality. We need to understand that we have a connection with the past and the future, and that there is a world beyond, which can't be reduced to words. It is absolutely fundamental to the human condition for us to express the things beyond our world in an artistic way. That brings us to something that is uniquely and universally human, like music. Something that guides us into the kingdom of language and poetry, toward the meaning of ritual and dance. And that all brings us together and connects us to our surroundings. Art is not what that horrible American critic calls cheesecake for the mind.

> You're talking about the psychologist Steven Pinker, who offers an explanation for the origins of art, comparing it with junk food. And who says that just like high-calorie and fatty foods stimulate the pleasure centers of the brain – because they were scarce and their consumption was very advantageous back when humans were hunter-gatherers – music, for example, elicits pleasure because, in our evolutionary past, understanding the sounds of nature would have been an advantage in terms of knowing what was going on around us, whether that meant escaping from a predator or finding fruit by tuning in to the birdsong.

Exactly. But unlike Pinker, I don't think artistic expression is superfluous. It is essential to our well-being, and I would argue that it is fundamental to our understanding of our place in the universe and in the world. And I think *Homage to the Elements* tells us that the elements don't just exist on the periodic table; they have their own form and resonance. We can translate them into other representations and, when it comes to color, we can look at them as though each of them has a different form of life. This revelation tells us that all the elements are interconnected, and that is absolutely fundamental because it tells us something about how we are interconnected with the world around us. And meditating on that fact necessarily implies a sense of loss, because given the way we inhabit the world today, we are in danger. It's a danger that we have created ourselves. Why? Part of the reason is that we no longer revere our connections with all other things.

The Italian philosopher Nuccio Ordine wrote a book about that idea of art: *The Usefulness of the Useless.*
When everything is reduced to its utilitarian quality, it's separate from the rest. In the end, these are all old arguments and not particularly original, but the reason they interest me is because I think about what I do. What is the meaning of artistic expression? I believe that it has to do with approaching things that can't be expressed in any other way, but that are essential to our lives, essential to understanding the world beyond words or beyond logical comprehension. In a prehistoric world, a man or a woman who went out and broke off a tree branch wasn't committing an arbitrary act. Their act was connected with the tree and with themselves because they understood that there was a deeper meaning for that tree's place in the world. When they broke it, it would have been because they needed a walking stick, or medicine. An even if were in a fit of rage, it was an expression of that rage, a kind of violation of the tree. Those people, whom we might call primitives, could then apologize to the tree or make an offering, for example. An offering to imaginary people who lived in the sky and who weren't separate from them or from the tree, and who were simply a metaphor for understanding that we're all interconnected.

There are tribes that use hallucinogenic substances to make those connections explicit. Do you think art can provoke something like those substances do?

When we look at our world, we create the narratives and the stories to explain it. We make assumptions about the nature of the reality around us. And art, contrary to popular belief, is a window onto reality. In that sense, if you choose to look through it or enter into it, your perception will be altered, just like it's altered when you take a drug. If you take a hallucinogenic drug, you may see an animal that no one else sees, and which the logical part of your brain will say isn't there. But because you see it, it is there. Many people have understood that this altered perception of reality may offer a deeper meaning to what reality is. And that's exactly what art does. It has a metaphorical function, it isn't literal. Even the animals painted in the Chauvet Cave, in the Ardèche region, 35,000 years ago, are continually acknowledging the form of the rock itself. So there was a relationship between the drawing, the hand of the person doing the drawing, the vision, the connection with the animals and how they are understood, and the form of the rock. Almost as though the animals came through the rock to be with us. It's also true that, at that time, we lived in an animal world, and not in a human one. Perhaps that gesture was a window, an admission that they were living in an animal world. But I'm also tempted to say that it implied an understanding that animals had an existence prior to human understanding. And that window made it possible for those men and women to live in that world because they understood that there were multiple narratives. What's more, since the drawings in Chauvet were done underground, in the dark, by firelight, there was necessarily an alteration of perception. If they had been done outside, in the sunlight, they would have been looked at in a different way. As for us, there is an urgent need for us to change the way we think. It's essential to our survival. Although as I get older I become more aware of the limitations of what I do, I think that art can have a similar effect to taking a drug, because it affects our consciousness.

A Welcoming
Space

Eulàlia Bosch

There are two ingredients for enriching our perception of the world: light and freedom. Light, whether it's a flickering flame in a cave or the floodlights in an Olympic stadium, reveals spaces in our minds or in the world that we have never seen before. And as humans, great imitators that we are, when we see new things, it makes us want to do things we haven't done before, which means that light serves as an inspiration. But to be able to give form to that inspiration, we need a space of freedom that doesn't inhibit our initiative. And freedom is a subtle thing, which often resides in elements that are inconspicuous and undervalued.

It's clear that if it has been forbidden to write that the Earth revolves around the sun, it will be difficult to offer an accurate description of the cosmos. We could go back 400 years and ask Galileo about that. It doesn't necessarily take a savage ban like that one to undermine the foundations of personal initiative, though. Arrogance can do a pretty good job of it on its own. Unfortunately, often enough, the same people who embody that light are the ones who obstruct the actions of others who have been inspired by their new and penetrating gaze. Who hasn't had an idea that they haven't dared to express in front of someone they admire, out of embarrassment or, directly, for fear of being belittled? Freedom, then, must be supported by a welcoming space, a comfortable place where the drive for novelty, innocence, and even error can serve as the jumping off point for a germination that offers free reign to the exaltation that comes from inspiration. Eulàlia Bosch has the virtue of combining the capacity for illumination with the offer of welcome. She has a flashlight that she uses, there can be no doubt, to shine a light on things you've never seen before. Plus, she has a dining room with comfortable sofas where you can drink hot tea and eat chocolate cookies, where you can pursue, without fear, the inspiration ignited by the vision of all those things. The combination of those two fundamental elements forms her view of the world and, in particular, of philosophy and education, which should always go together. A kind of education that should help people grow individually – not like an isolated palm tree in an oasis but like a tree in the midst of an ancient forest, connected by branches, leaves, vines and roots to everything around it in space and to everything that has come before it in time. Community and legacy. That's just one of the combinations Eulalia Bosch shines her light on.

How did you meet Eugènia Balcells?

By chance. I was walking down the steps of the La Virreina with a friend, and we were talking about Balcells' piece *On and On*, which we had just seen. The visit had sparked a lot of comments about the magic it exuded and the artist's generosity in letting the viewers see the mechanism that generates the images, which in fact, uncovered the meaning of the piece. Suddenly, she stuck her head out. My friend, who knew her, said, "You'd be interested in one another." We talked for a bit about what we were doing and, following up on our mutual friend's intuition, we decided that I would visit her studio the following week. That was in 1994.

You've worked with her consistently since then, which also made you one of the few people who was there when *Homage to the Elements* was born. How was the piece conceived?

Homage to the Elements came about as the result of a desire to accompany visitors – and, most especially, teachers – through the installation *Frequencies*, which was shown at Arts Santa Mònica in Barcelona in 2009 . We felt that reminding people of the periodic table and showing them that each element has its own light signature was important in helping people delve into the contemplation of the piece, a projection of the elements' emission spectra that is just hypnotic. So we made a poster of the periodic table with the light codes for each element, and we hung it in the visitors' area.

So it was born as a kind of explanatory key.

In the years that I've been a teacher, I've seen many examples of how explanatory formats that teachers invent to help children in early learning and primary school to understand certain concepts end up being incredibly interesting to adults, regardless of how they originated. The case of *Homage to the Elements* is perhaps the most extreme. The first poster was made with schools in mind, and immediately it sparked a huge interest among the most distinguished professors from the Faculty of Physics and Chemistry at the University of Barcelona, who tirelessly pursued the edition of the magnificent mural we have today.

There are processes of synthesis in both art and science, and *Homage to the Elements* is a clear example, in the sense that it crystallizes a certain natural order. At the same time, would you say that, in a way, the mural is a synthesis of Balcells' career?

Homage to the Elements is a moment of synthesis that comes in the wake of a long-running fascination with light and its ability to breathe life into matter. We shouldn't forget that film was the starting point for Balcells' artistic production and that light has been, at the same time, the object of her research and the main instrument in her installations, and even her graphic work. But what I find most curious is that when *Homage to the Elements* was made into a large mural and it began to be installed in universities and schools, it seemed like it was the initial motor that had launched Balcells' career as an artist. It was as though, all of a sudden, we had forgotten that it came after essential installations like *A Space of One's Own*, *In the Heart of Things*, *Transcending Limits*, *Exposure Time*, *TV Weave*, *Dress of Light*, *From the Center*, or *Wheel of Time*, and so many other pieces in which light is the indisputable center point.

What do you see in *Homage to the Elements?*

I see a toolbox, which, like the one a plumber might carry, has two overlapping levels. One holds the energetic capacity of the sun as the origin of life, and the other contains the complete history of painting: I see the red and black of the bison in the Cave of Altamira, the paintings from the houses in Pompeii, Giotto's frescoes and Rothko's canvases, Antonio Saura's black palette and Miquel Barceló's still life colors, and the golden color of Wolfgang Laib's pollen... I see them all, especially when I look at the elements' spectra in motion in the installation *Frequencies*, and the screen is filled with pure colors that chase each other around. I remember how, during the exhibition, we explained to the children in early education and primary school, who were curious about the images in *Frequencies*, that *Homage to the Elements* is the box of crayons that the sun uses to paint the world each morning. In doing so, we also reminded them of the darkness of night, even though it has long since disappeared from cities...

You've spent many years thinking about how to teach philosophy and doing it in many different ways.
What is philosophy for you?

I've always complained about my education because it was founded on specific questions in specific subjects that had unequivocal answers. In that sense, everything was technical or based exclusively on memorization. It was only when I finally took History of Philosophy at the end of high school that I realized I was interested in continuing down that path. My fundamental critique of school was that no one ever told me what the different branches of knowledge were really about: chemistry, physics, math, literature... How did I pass? I can't tell you. I really can't. If I had known what the different subjects were about at the right time, I would have been able to make an informed choice, but instead I just ruled out anything that came with a solution key.

Wasn't that the case for philosophy?

I clung to it because I understood that it was an ongoing effort at questioning things, which let me prolong the admiration I felt from discovering the world and explore paths of reflection that had the potential to go on forever. It was as though, as I walked along, new doors and windows kept opening that I needed to explore. I have to admit that I have a recurring dream in which I discover a room in my house that I didn't know was there, and I keep asking myself how it was possible that I had never seen it before.

And today, philosophy, for me, is still that attempt to construct a space to live in based on the questions that life offers up to me. I never set out to pursue an academic career. I've always been interested in trying to understand everyday life within the universal whole that gives it meaning. And I like to quote Juan Claudio Cifuentes – Cifu, they called him – who would close out his radio program *A todo jazz* by letting the last echoes of the saxophones, trumpets or drums, die out completely, and once there was complete silence, he would say: "On this show, we respect the reverberations..."

Keeping those reverberations going, the drawn-out duration of little everyday events, is the best stimulus for reflection, reading, travel...

In that sense, my interest in philosophy made it essential for me to know what the sciences and the arts are all about, and now I find they're inseparable from one another. Just as it was in ancient times, philosophy, the effort to find meaning in life, is still the core around which all knowledge grows. And poetry is its great ally.

What is philosophical about *Homage to the Elements*? Because of my involvement with Eugenia's exhibitions, I've spent many hours in front of *Homage to the Elements*, especially when *Light Years* – the exhibition it was part of, along with *Frequencies* and *Universe* – was traveling around. And contemplating the mural always takes me into the terrain that the pre-Socratic philosophers began cultivating. How could they be so exquisite in their wonderings about the origins of reality, as they walked the tightrope between the mythical world of the gods and the very first scientific reflections without any of the instruments we consider essential today? How could Lucretius talk about atoms colliding and compare them with letters combining to form words with such diverse meanings, based only on his sensory perceptions and an inherited language? Or so many centuries before him, who could have written in the *Rigveda*, the sacred verses of Hinduism, the text that reads: "Darkness there was at first, by darkness hidden/ Without distinctive marks, this all was water/ That which, becoming, by the void was covered/ That One by force of heat came into being"?

Homage to the Elements makes me think so directly about those moments from the early days of humanity that I chose fragments from the writings of those ancient sages to accompany the mural in the exhibitions. I was aware that I was adding the difficulty of poetry to the surprise with which people received the mural and the installations *Frequencies* and *Universe*. But the great virtue of poetry is that it offers moments of intense light and, at the same time, it maintains the mystery which is, at its core, the seed, the nucleus around which human knowledge revolves. I wish poetry played a more important role in our lives! I wish we had learned more poems by heart at school! Maybe then we would look at *Homage to the Elements* as a sign that is both scientific and magical at the same time...

> Your vision of poetry reminds me of how Jorge
> Wagensberg said that there are three forms of
> knowledge: art, science, and revelation.

There can be no doubt that science is looking for the answers to
fundamental questions. The arts, when they're genuine, do the same
thing. There is no doubt in my mind that revelation exists. It comes
on suddenly and takes you somewhere you didn't know you were
going, whether it's in the broad realm of the arts or the vast universe
of science. When I write, it's like a dictation. I take down what I'm
seeing, what I've heard, what I think I've perceived. And I write
by hand, without stopping, without going back. I write like it's an
unstoppable impulse, and suddenly I'm moving through sentences
that describe unknown universes and that demand that I keep going,
going, going, letting my hand continue, even faster than the mental
constructions of the words. And, later, when I go back to those pages,
I'll find an image that I can't believe I wrote. As though someone,
neither entirely inside or outside of me, were guiding me in the
writing, as is sometimes the case with thoughts on a subject that may
be enormously complex, but which suddenly arrive in a place that is
entirely clear and open. That's why I always need a piece of paper and
a pencil within reach, so that I don't miss that moment of revelation
when it comes on suddenly.

> You've worked for many years to introduce art into
> schools. Why do you think it's so important?

Art keeps the spirit of questioning alive, and it prevents the division
and categorization of the different branches of knowledge at a time
when that division isn't relevant at all. Just the opposite. The emotion
that comes from an aesthetic experience – and learning has a huge
and unquestionable aesthetic component – is a reaction of the person
as a whole: from head to toe. Early education and elementary school
children have a global notion of knowledge. They learn, and that's
that. And they respond to the arts based on that indivisible unity, and
which the arts actually demand but don't always find among adults.
On the other hand, introducing artistic experiences into school groups
has the added value of incorporating spoken dialogue, which gathers
the reactions from different people brought on by the same music, the

same painting, play, dance performance or film into a conversation, which offers an outstanding learning environment. Then, despite the difficulty associated with it, constructing an argument becomes attractive. Hearing yourself speak and being heard is one of the most powerful qualities of humanity in a human being, and the arts let us put that into practice with unusual ease. Like teaching, although I say that parenthetically.

> Since you brought up teaching, there's an intense debate going on right now about how schools, and education as a whole, can best prepare children for an unpredictable future. What role should schools play in this situation?

First of all, school is a new space. And you have to leave the nest to get there. In general, it's a process that involves a lot of tears. Yet, once that unfamiliar space becomes a comfortable place, it can be the source and the destination for any number of excursions that help you to understand your surroundings. Surroundings which, over the years, can take on an extraordinary scope.

Going on a school trip to learn about the neighborhood and its library, going to the theater, to the circus or a museum, going to visit a hospital or a post office, a workshop or a factory, a special shop or a department store, to attend a concert, a dance performance, or to watch a film, to visit another school or to go to the beach, the botanical garden or a construction site... And then going back to school to follow up on all those stimuli with the help of teachers, classmates, the school library, the internet... That's one of school's essential purposes. It's also an achievement that requires an almost infinite supply of energy – which only teachers are able to sustain – but which often extends far beyond the expectations and provides heaps of knowledge and well-being that reverberate for a very long time. When that happens, horizons are broadened as the years go by, and finishing school should mean being fully situated in the community in which the students have grown up.

> In the efforts toward achieving that goal, there's a debate going confronting so-called old and new pedagogies.

I don't think schooling should be uniform. However, it should be public, and in public schools everyone should be able to experience all the things that we believe can nourish the terrain that each boy and girl comes in with as brand new. That doesn't have to mean that everyone does the same things everywhere, but there has to be respect for the right to fully understand all the heritage that humanity has accumulated so far.

The aim of schools is to awaken the spirit of questioning, to share it and, using the means available today, to try to satisfy it as far as possible. In that sense, the presence of the philosophical spirit in schools seems essential to me. Much more important than the concrete forms it may take in each case.

I also think it's very important for towns and cities to recognize their responsibility to collaborate with schools and for institutions to be willing to play their part in the education of the new generations. Schools aren't the only places education can take place, because if that were the case, education simply wouldn't fit into them. I've always liked to imagine school hallways as a natural continuation of the streets. Making everyone who should be implicated responsible for education is far more important than offering an hour more or less of any one subject.

> Would you like *Homage to the Elements* to
> be in schools?

If it were up to me, it would be hanging at the entrance to every school in the world. I've been working toward that for years, and this book is an effort in that direction. I feel a strong connection between that desire and the sustenance of human life on Earth because *Homage to the Elements* expresses, at the same time, what matter is made of and how far humanity's knowledge of it has come. In that sense, it contains everything we know and everything we need to put into motion to keep life going and to make it permanently renewable and sustainable.

For years, I've been imagining *Homage to the Elements* as a bright, magnificent mural in the hallways of schools around the world, welcoming families every morning. I imagine it like a camera that captures the moment when the adults and the children go their separate ways, an act they carry out for the sake of the transmission of knowledge. And maybe one day a parent or a group of mothers talking at the entrance to the school will wonder what the mural means, and they'll share their curiosity with one of the teachers. At that point any of the teachers will be able to give their version, talking about it from the standpoint of physics, chemistry, physical education, medicine, philosophy, history, poetry... and all the explanations will be true at the same time. That's a wonderful image, don't you think?

As for the girls and boys, there will come a day in their schooling when Mendeleev's regular table will be the subject of one of their lessons. When that happens, they will already have internalized the image years earlier, and everything will start to take on a new life inside them. They'll start to understand why an artist needed to pay homage to the protagonists of life through an ode to the common origins of everything that exists.

I'd also like to think that contemplating *Homage to the Elements* in a group and with teachers nearby can awaken a strong sentiment in favor of peace and humility, which is so often at odds with much of human behavior today.

I would love for *Homage to the Elements* to be a distinguishing feature of every school, and for all the children in the world to finish their schooling having shared in the experience of light that comes from playing in front of the projection of *Frequencies*– as we saw when the exhibition *Light Years* traveled around the world – and knowing by heart the passage where Plato tells us that knowledge should be used to understand how we want to live.

On the Surface of Things

Sally Potter

When Sally Potter told me that film was about showing the surface of things to explore what they're like inside, it took me by surprise. And for two reasons. First, because I think these kinds of short, seemingly simple phrases that can only be arrived at after years of experience are like intellectual diamonds. This distilled but defining vision of cinema can transform the way we watch and think about films, and that shift can cause a leap forward in our aesthetic appreciation of art and even of the world. Second, that way of exploring reality, which results, in film, from the limitations of cinematic language, is very similar to what scientists do when they're studying natural phenomena. When the British scientist Ernest Rutherford discovered the atomic nucleus in 1913, he did it by bombarding gold foil with alpha particles (consisting of two protons and two neutrons) and observing how they were deflected. In other words, the analysis of how the particles bounced off the surface of the gold foil revealed to the researchers the existence of a structure inside the atoms. Sally Potter wasn't expecting her notion, which she had internalized to such an extent, to spark another idea or association. The paths of knowledge, however, are inscrutable. So are the ones opened up by her films, an intense and evocative aesthetic experience that offers new readings about the human world and a certain distillation of the human experience. Who would have imagined that science and film operated in such a similar ways?

> What did you think of *Homage to the Elements* the
> first time you saw the mural?

The first thing that came to mind was *The Periodic Table*, the book
by Primo Levi. I read it decades ago and loved it. Reading about
the fact that the world – everything, actually – can be divided into
the chemical elements that make it up offers a kind of philosophical
satisfaction. And sometimes I refer to that satisfaction in my own
work. But what Eugènia Balcells does diverges entirely from that.
Because she translates the idea that everything is made from the
elements into a representation using the signatures of light and certain
equivalences in sound. And it almost becomes an artistic vision of the
theory of everything. I think it has that kind of ambition. And also to
get to the root of what we can see of ourselves and what we're made
of. It's a very ambitious and exciting artistic project. Another very
interesting thing is that the piece's appearance is deceptively simple,
and at the same time strangely familiar, because of the idea we all have
in mind of barcodes. It's like it came out of the supermarket of life. All
the elements together in a combination where there's abstraction, an
economy of means, and simplicity. It's fascinating to look at.

> Balcells works with light as though it were almost
> a plastic material. In film, light is a fundamental
> element.

It's absolutely essential to shooting a film. In fact, films would be
unimaginable without it; light is the foundation. You can make films
without sound, although there is always the ambient noise of the
world as you're watching them, but you can't have a film without light.
Films can be made using analog or digital media, with chemical or
electronic technologies, but you need light to record and to project.
Film is an intangible medium that exists only through light. That's a
formal foundation, but then, from a practical point of view, you also
have to work with lights. For example, if you're shooting at night, you
need to use artificial lighting. When you're shooting a film, you're
always working with an awareness of how light changes the space,
people's faces, and the geography in general. There's a lot of discussion
and experimentation with light during a shoot. Different light sources
are set up, and there's a team of people who work on it. All of this has

to do with how human beings consider the plasticity of light in film and its practical use in recording the surface of things to explore the non-materiality of everything. Film is a projection that you can't touch or smell or grasp. It's always in motion. And if we're talking about the fundamentals of film, the other one is time. Film is light and time. The rest is secondary.

> You've been quoted as saying that people are all the
> same on the inside, whether they're black or white,
> Jewish, Christian or Muslim. And that's easy to say,
> but it's hard for many people to accept. Do you think
> that works like *Homage to the Elements*, which offer
> a universal view of the world and of humanity, can
> help support that vision of other people?

I would say yes and no. Anything that reminds us that, fundamentally, as human beings, we're all made of the same material, like blood — our blood is all the same color — everything that reminds us of what we have in common rather than what makes us different is incredibly helpful. That being said, an exhibition can be held in a gallery, or a school, or a university, for example. And those places are associated with a specific target audience or a social group that commonly frequents that type of space. In other words, there's a kind of social and political map into which this piece – which is so universal – is incorporated, such that, at the outset, people's expectations may be very different. It's very interesting for a work of art to cross those barriers of expectations and transcend them. But even if it can't, in each case it's still acting as a reminder of something universal.

> You've also said that freeing ourselves from the past
> and seeing the world with fresh eyes, like children
> do, can help us experience the present more intensely.
> Can this kind of artwork help us achieve that state?

Everything that offers a surprise or a new way of looking – what has been called the shock of the new – drops us into the present. *Homage to the Elements* isn't specific to a particular historical period; it goes beyond that because the light emitted by the elements is the same now as it was in the Middle Ages, or in ancient Greece, or even before humans existed. The principle of the piece was already there in those days, but it wasn't visible. In the past there might have been similar visions that drew on the language of the gods, but not on chemistry. But in terms of emerging from the past and arriving in the present and how the piece achieves that, I would say that the discoveries it contains and the manifestation of light in relation to the chemical elements are new things. So it appeals to the present moment, it somehow takes us to a place where science and art have come together, but in this very moment, not in the past century or many centuries ago, not even a few years ago, but right now.

> Every work of art says something, and I'm not
> referring to a moral message, but to the fact that
> artists are, in theory, people who have something
> to say...

Not necessarily.

> No?

Artists don't necessarily have to have something to say, it's more like they have something to find. They have to find what they're trying to say. There isn't necessarily a predetermined message – whether moral, or political, or of any kind. And if there is, they're making a mistake. The work won't be alive because it won't be a process of discovery. Pride or disappointment can lead artists to claim that they know what they want to say, but I've never met a good artist who contends that their art is anything more than a form of exploration. You could use the term research, or investigation, or invention, but the most common one is discovery. In the case of *Homage to the Elements*, it's about finding something that already exists. Balcells found something that has always existed, but she isn't saying anything about it. She's showing us what she found. And that's something more like being an archaeologist of phenomena.

> Artists have access to lots of different mediums for
> their explorations. Are there aspects of reality that can
> be explored by writing a book that can't be explored
> by making a film? I say this because in the process of

shooting *Orlando*, I understand that you tried to get
to the essence of Virginia Woolf's book and to express
that proximity through the language of film.

Each medium has its own specificity. Every artistic language has its
strengths, its potential, its limitations, and its glories or triumphs.
There may be a process of translation because there are connections,
parallels, replications, or reproductions. But, above all, there is this idea
of translation. The analogy of translating between spoken languages
is a good one. Are there things I can say in English that I can't say in
French? I think there are. A different language structures our thinking
and what we can express in a different way. Plus, language is imprecise,
it's a kind of education from the field of thought that we hope can be
transmitted through the relative precision of words.

What, specifically, can be explored in a film that
couldn't be explored in other languages?

Film is a hybrid medium; it's very impure and, at the same time, very
synthesizing because it combines many different elements together.
As I said before, film works with light and time. In my case, when
I make fiction, I also work with the written word, with a script that
is the architecture of the film. But then you also have to work with
actors, spaces, places, times of day, narrative, sound... There are a lot of
elements of sound – a film can have 50 tracks! – and then there's the
music, of course, if there is any. All of that makes film a very complex
medium which, nevertheless, has to create content that we can relate
to in a very direct way. And that creates an apparent simplicity, unless
what the author wants is to create confusion. But if, as an author,
you're aiming for some kind of clarity, you have to manage all those
many ways of communicating something. I've been working a lot with
music recently, and I've realized that there are some things you just
can't do with music. You can't use it to represent a narrative idea, for
example. Music evokes a world of abstraction and emotions based on
mathematics. That's its beauty, but it isn't a language for telling stories.
Even songs with lyrics can't tell a story the same way film can. Film
is a seemingly representative medium because it shows the surface
of things with the goal of evoking the non-material realities that lie
behind that visual and familiar existence. In addition, because of its

complexity and its representative nature, film can deal with many parallel levels of reality and communicate them all at once. In truth, because of all that, it's more like a mirror of consciousness. I always think of it as a projection of the mind.

> How would you imagine a translation of *Homage to*
> *the Elements* into the language of film?

It would depend a lot on the type of film. The films that Balcells usually makes have an abstract style that is closer to video art, less narrative, and from that perspective there would be many ways to explore it. If it were a fiction film, there would have to be people and a story, in which the elements would play a role. The people might interact with them and work with them or discover them. But it would be something very different because *Homage to the Elements* is a work of art designed to be perceived in a very specific way, through the language of abstraction and the desire to not represent anything more than that barcode, representing itself and nothing else. And yet, the piece evokes other things, like how we feel about the fundamental architecture of the universe, for example. In that sense, it offers an opportunity for contemplation.

> Making a film entails working with actors, who
> are human, so there is always a certain degree
> of unpredictability, something that is outside
> your control.

That lack of control is part of the process of working in film. And it's not just because of the actors. Nothing can be entirely under your control when you're making a film. Not time, not what happens behind the camera, not the budget... It's a very volatile process. As a filmmaker, you can only do the work if you're prepared to somehow ride the dragon of that volatility and unpredictability. In the end, it's a question of redirecting and incorporating that unpredictability instead of controlling it, ensuring that all those unpredictable elements are flowing in the same direction. If they're going in different directions, you don't have a film, you have a mess. However, if they're flowing in the same direction, they can enrich the author's original vision. That said, the issue of control and being open to incorporating volatility is always a dilemma for the director.

> In *Rage* you drilled down to the essence of film: an
> actor, a frame and a color.

True. Plus, it was also the first film in history to be released for mobile phones, which conditioned the approach and the elements we used. In fact, it was a little too early. When it premiered in 2009, there were only some tens of thousands of people who had phones that were equipped for streaming. Five years later there would have been millions, but at the time that was a prophecy rather than a reality. People were telling me that it would never happen, that there would never be anyone watching anything on their phones.

> So is that the essence of film: an actor, a frame
> and a color?

That's part of the essence. When I was designing *Rage*, I was thinking of a thumbnail, though. It might not be very interesting to watch *Lawrence of Arabia* on a mobile phone, but I asked myself what it might be possible to watch on a phone and what people were watching at the time. And what people were watching were their friends – that is, a face in a frame. So I thought I would take that language and make something with it. I worked on creating an imaginary that would respond to how people use their mobile phones, how they move them when they're looking at them, and so on, so it could be translated to the camera. In any case, it's fiction: we want to frame other people's faces and look at them, hear what they're saying, understand them, study them, but not in a mutual way. That's the big difference with what we're experiencing now, in our mutual interactions where screens act as mediators. *Rage* was constructed as a unidirectional gaze.

> The film, like *Homage to the Elements*, has a
> deceptively simple appearance. It seems easy to
> do. There are people who might wonder what the
> difference is between these more essential kinds of
> films and the amateur videos people record at home
> and post on social media.

The line has been blurred. But for starters, a movie like *Rage*, for example, features a dozen of the best actors in the world, which isn't the case with a YouTube video. And that entails a lot of work in terms of seduction, organizing, rehearsing, putting together wardrobe and makeup. The background colors, for example, were taken from something on the actors' faces or bodies. In one case, we used a lipstick color for the background. There's a lot of manipulation of those simple elements, a lot of writing and rewriting, structuring and editing.

In contrast, the language of YouTube is more like a glorification of amateurism. The mistakes and the instability of the camera, among other things, create the illusion of a kind of intimacy, of humanity even, thanks to the fact of removing the filmmaker's touch, which always structures and perfects everything in a very particular way. And that has had a huge impact on the syntax of film and how we make films, and especially television, which is another parallel universe. The way people use YouTube, videos by influencers, WhatsApp... It has all had an enormous aesthetic effect on cinema.

Can that influence enrich cinema, or does it trivialize it?
In truth, most of the videos you watch on YouTube are trying to imitate film. The fact is that making films has gotten a lot cheaper; anyone can pick up a phone and make one. When I started out, even making a short film was very expensive because you still had to pay the lab to develop it. Now, the fact that the means of production have become more accessible raises a series of fundamental questions: What should we be filming? What is interesting? What makes making a film worthwhile? Why will people want to watch it? On the other hand, with so much content, so many images and so many videos circulating now, it's like a sugar high. And that excess leads to a feeling of apathy, a soporific sensation. But that's good for filmmakers because there's so much content that, as a filmmaker, you can think, okay, what is really worth doing? What is fundamental? And that's a really good question to ask ourselves.

And what's the answer?
Oh, I don't know. But I've probably always thought that it wasn't worth doing something unless I really felt inwardly that I had to do it. I haven't always known why, but I've often felt like I've just *had* to do certain things. And it's taken a lot of passion, energy and commitment to carry them through. In that sense, making films isn't a normal job.

It takes a lot of work every day, that's for sure, but there has to be that drive, because otherwise, what's the point?

> Aside from that passion and doing things that have a point, when you're making a film now, are you thinking about how it can capture people's attention given the sheer amount of content out there, or do you just do it?

The most important thing is to focus on the film and believe in it. To look for what you need, find it, and do whatever it takes for the film to be made. However, as a filmmaker, you also have to be in contact with the real world. If you make a film and no one wants to watch it, you end up in a depressing situation. And sometimes that happens. There are films that get a lot of exposure and that are seen by millions of people, and there are others that are made with the same amount of energy that just fade away. Although some of them may reemerge later. But thinking about how to get people's attention is for the marketing department. And that doesn't really appeal to me. For me, above all, I want to be making something that I would go to see. I figure that if I would go to watch it, other people will too. And not necessarily because it's related to a personal interest of mine, but because there's something in it that seems necessary. On the other hand, as a filmmaker, you do think about what kind of world your film will be landing in. What are other filmmakers doing? What are people going to see? All that has become very complicated with the pandemic because many cinemas have closed, and the health measures have made it very difficult to shoot. Material conditions have a big effect on the medium, which, naturally, doesn't exist in isolation from the world.

> Some of your films, like Balcells' works, aren't situated in the middle of the market. And that can be risky, can't it? In fact, you took out a second mortgage on your house to shoot *Rage*. How do you manage that lifestyle?

I'm used to it because I've been working in art full-time since I was seventeen. I've never had security; I've never had what a lot of people would call a proper job. I've never had a steady income, I've always travelled around, and that's very useful when it comes to picking up a kind of survival skill that involves being able to tolerate insecurity and an unknown future. I think that's absolutely necessary if you want to get involved in art. On the other hand, no one really knows what the middle of the market is. Big blockbusters buy their place in the middle of the market, but they often fail miserably, and there are some that no one goes to see, so there's no recipe. And then there are some small productions that stand out, like *Orlando*. Nobody had heard of me when I did it, and yet it managed to do better than some of the blockbusters on the lists. We live in a world of surprises. But there is a time when, as Balcells probably does, you accept your status as an outsider, a commentator, not being at the commercial center of culture, although you may be in a different kind of center of a different culture.

And does that pay off?
It doesn't pay off with money, but it does pay off with a kind of self-satisfaction in pursuing things that you believe in strongly or that you feel are necessary. Also, because you build a body of work that sometimes has an influence in ways you hadn't imagined. Two years ago I was in Poland, where they were doing a retrospective of my films. After a screening, almost the entire room lined up to talk to me individually and to tell me what my films had meant to them. And I didn't even know that people were watching them in Poland. At times like that, you realize that as a filmmaker, or a writer, or an artist, when you release your work into the world, you never know where it will land. And when you find, unexpectedly and maybe years later, that they've landed right in the heart or soul of someone else, a stranger, it's miraculous. Then you realize that even if you're making films that you'd like to watch yourself, you're actually making them for strangers. Strangers that you love. And that's very moving.

Cave Performances

Roberto Ontañón

The first time I stood in line for the Chauvet Cave, I felt a weight and a tingling in my stomach. The typical butterflies? It wasn't just that. Although visitors can only access a reproduction of the space, the feeling I got just before entering was a mixture of joy, reverence, and a sense of connection to something deeply and essentially human, which must be related to the symbolic thought embodied in the paintings done in caves tens of thousands of years ago. The people who painted them lived in a very different way, and from our place in the present they may seem very far away, but their brains, and therefore their creative impulses, were the same as ours. Once inside, that mix of emotions quickly dissipated because my 10-month-old son, whom I was carrying in a backpack, spent the entire visit crying. I didn't hear any of the explanations given by the guide, an individual whose white beard advertised a level of experience that made me want to pepper him with all sorts of comments and questions. That day he only looked at me occasionally with reproach because of the noise my son was making, but I would be returning to Chauvet. And I would also visit Lascaux and Altamira, and other caves. Every time I enter one of them, the joy, the reverence, and the sense of connection are still there. That's why I'm a little jealous of people like Roberto Ontañón, whose job it is to visit the original caves and study those magnificent paintings to find out a little more about them every day. Every discovery they make, every article they write, and everything they explain to us brings us one step closer to the fascinating minds of the people from those caves, which, ultimately, are also our own.

What is your impression of Eugènia Balcells' work?
Her work extends over a very long stretch of time and offers many
different approaches to the artistic phenomenon. It's very multifaceted,
and it delves into the connection between art and science, something
I find very interesting and which doesn't get a lot of attention, at least
in the academic world. I've also been very interested in her video
pieces, because of the way she explains things, and her installations,
because they tie in with some of our current approaches to rock art.
Although this may seem like a cliché, I think Eugènia Balcells is like
a figure from the Renaissance, because she has a lot of interests related
to many different areas of reality and knowledge, beyond the popular
conception of art as an aesthetic pursuit.

> Does the treatment of light in works like *Frequencies*
> or *Homage to the Elements* have anything to do with
> what painters were doing tens of thousands of years
> ago in caves?

It makes me think about it from two angles. First of all, light is an
essential part of Paleolithic art, at least the art that was done in caves,
because it was carried out in the deepest areas of the caves, where
there is no light at all. When you try to engage in any activity in a
completely dark place – not just an artistic activity – it is determined
by the light. In cases like the ones in some caves in the Dordogne,
or the Cullalvera cave in Cantabria, people ventured as far as 1,200
meters into certain caves, with everything that entails in terms of
exploration, speleology, and feeling along the walls for orientation.
And they had to be doing it with the help of artificial lighting. In that
sense, they needed a mastery of light that would allow them, first, to
get to such remote places, and then use that light to work with the
forms of the walls and the ceilings of those caves. And that last aspect
is fundamental, because in the very beginning of artistic manifestations
it wasn't an activity that was being done on a blank sheet of paper or a
flat canvas – on the contrary. The first artists created forms using light.
I have a Belgian colleague, Marc Groenen, who wrote a book called
Sombra y luz en el arte paleolítico [Shadow and Light in Paleolithic Art],
in which he argues that the paintings weren't so much a question of
lines or tracings but of light projected across the natural forms of the

caves. What those people saw in their daily lives completely immersed in nature suggested animals, humans and mineral shapes to draw, but the conditions were very different inside a cave. The lighting was dim. They used oil lamps that gave off a kind of light that was more or less like candlelight. The lighting system is already interesting in itself because it isn't based on a fixed light source with a single frequency, it's an errant light source, in movement. And that kind of light, used in an irregular environment like the walls and ceilings of the caves, allowed the first artists to shape the figures using those natural forms, rounding them out with lines or engravings. Their light, like in Balcells' work, is creative in itself. It isn't a complement; it is part of the art of the Upper Paleolithic.

On the other hand, *Homage to the Elements* reminds me of a certain aspect of our research into rock art. We work with spectra to analyze the pigments in the different figures, so we can compare them. We use the spectra reflected by the paintings when we shine a laser on them to identify which pigments were used. In other words, Balcells' work is reminiscent of both the object of our knowledge and the methodology we use to study it.

> In some caves, like Chauvet in France, there are a series of superimposed figures of the same animal which, under the flickering light of those oil lamps, generated a certain sensation of movement. Could that be considered a kind of protocinema?

The originator of that interpretation is a French researcher named Marc Azéma, who published a book called *La Préhistoire du cinéma* [The Prehistory of Cinema]. Rock art has always been understood as a static thing, as a series of figures in hieratic poses, without a background or a frame, that aren't set within a landscape. In reality, however, rock art is a much more complex production. At various sites, elements have been found in the paintings and bas-reliefs that lend them a dynamism, like the duplication of legs or heads, for example. There are even figures that appear to use natural reliefs like stalagmites to reproduce a trotting movement. There are others that look like they're raising and lowering their heads or turning them left and right. And how were they set in motion? How did Paleolithic artists make the drawings on the walls start to move and engage in a kind of performance? With light. They would walk along the wall with a lamp in their hand or move it from right to left or up and down. And that made the figures come to life. When you see it, it's spectacular because you feel like the figures are really moving. And this ties in with what I was saying earlier about rock art, that it isn't just a painting on a canvas or a bas-relief on an Assyrian staircase or a Greek pediment. Rock art can be understood as a real artistic installation that relies on the participation not only of the authors and the works of art, but the spectators too.

Does that mean that people in the Paleolithic went to see them like we'd go to see a film today?

Recently, along with a Japanese colleague, we published an article that looks at rock art like an installation art. In the paper, we looked at how some figures of animals from different caves in Cantabria are drawn in seemingly disproportionate or distorted ways. But the thing is, depending on how you look at them, you see them in one way or another. The artists used the rock to situate the figures in such a way that favored certain perspectives. In this case, the figures weren't imagined from the standpoint of a value that was derived from their execution, as is the case with movable objects; instead, the meaningful moment was when they were being observed. We've also analyzed how they are situated on their supports, and we've noticed that some figures are located in places that are designed to be seen by a large number of people, because they're situated in areas where various sightlines converge. But there are others that are in places where just one point of view lets you see the figure with the right proportions. Those figures seem distorted when you look at them from a different spot. They look like they're poorly made, but it's an effect that is sought out intentionally, not an inaccuracy on the part of the artist. And that ties in with the work of contemporary artists like Balcells, who has created montages that play with volumes, with spaces and with geometry.

Pablo Picasso once said that the best painter in history had painted in a cave tens of thousands of years ago, but on the other hand, this type of art

installation fell out of use for centuries. Could we say
that Paleolithic art was ahead of its time?

I don't think it was ahead of its time, rather that we tend to
underestimate prehistoric people. We assume that, because of
their lifestyle, as hunter-gatherers who were very attached to their
ecosystem, they must have had a limited capacity for abstraction
and underdeveloped skills. They were people who worked with their
hands, and very expertly. The objects they produced, which are a very
important part of Paleolithic art, are the result of millions of years
of working with bones and stones. And some curious natural shapes
began to appear in those objects that captured their attention. And
here we're not talking about *Homo sapiens*, our species, but about our
ancestors – some of them very distant. There were colorful objects and
objects with shapes that caught their eye, but in those objects, mostly
stones and bones, we also find incisions and rhythmic indentations.
And later, those incisions and dots begin to be seen on walls, where
the dots form lines, and all that ties in with the configuration of rock
art. In any case, they were people with tremendous amounts of skill.
There are many pieces in which it's obvious that the criteria for their
construction were not purely functional, because you can discern
in them a conception of size, symmetry and color. One very clear
example of this is the famous "Excalibur" found in Atapuerca: a red
quartzite hand axe crafted by people 400,000 years ago, before the
Neanderthals, probably from the species *Homo heidelbergensis*. In the
objects those people made, it's clear that they weren't working with
strictly functional criteria, but that there was already a certain aesthetic
intent. Put simply, they liked to make pretty things. And as we move
forward in time, we also see this in the early emergence of the mobile
art made by *Homo sapiens* in Europe 40,000 years ago. There are a
series of small figures shaped like animals that are absolutely exquisite.
And they are among the earliest known figures, even before rock art.
These people were prehistoric, but they weren't primitive. We need to
separate those two concepts.

From a cognitive standpoint, then, were they like us?

The way they execute rock art is entirely unrelated to children's art
or what children do, a parallel that has been pursued by certain

psychologists and anthropologists who have studied the origins of art. From an artistic point of view, Paleolithic art is superlative. From a technical point of view, they were able to use the materials at their disposal to make the best tools for their survival – proof of which is that fact that we exist here today. At the same time, this technical skill was combined with a capacity for abstraction and expressing ideas through symbols that is exactly the same as ours. Starting 200,000 or 250,000 years ago, which corresponds to the oldest remains that have been found of *Homo sapiens*, these people had a neurological configuration that was exactly the same as ours. All their capabilities were there. The rest was just a matter of trial and error, of trying out new materials and developing technologies. By using those technologies they were able to change their way of life and their capacity for survival. They ended up becoming sedentary and creating settlements and civilizations, and that, fast-forwarding, brings us to where we are today. Ultimately, the physical constitution and the brains of *Homo sapiens* from 50,000 years ago are indistinguishable from our own. In that sense, we shouldn't be surprised that from both a technical and an ideological point of view, they were exactly the same as us. I insist: they were prehistoric but not primitive.

Why, then, did Paleolithic art give way to simpler art forms rather than evolving into more complex ones? The kinds of artistic production that are disconnected from direct subsistence tend to fluctuate. Throughout history there have been different needs and ideological foundations. When you compare classical art with Romanesque art, that becomes very clear. We often hear about the pinnacle of Greco-Roman art, a technical art characterized by perfect proportions, and, in contrast, Romanesque art, which came much later, seems coarser. In Romanesque art the perspectives seem to be flattened, the compositions look simpler, but in reality the styles represent two different worlds, different cultures. They are different responses. In Paleolithic art, there is a culmination in the Magdalenian, with the polychrome figures from Altamira and Lascaux, and the incredible black figures from Chauvet. If we keep moving forward in time until 10,000 years ago, however, the glacial period is over, it's the start of the current epoch, the Holocene, and the same people, the same human groups start living in a different way. The Paleolithic is over, the Neolithic is beginning, and art – what we call art, both the decoration of small objects and rock art – undergoes a fundamental shift. Two successive cycles emerge in the Iberian Peninsula: Levantine rock art, characterized mainly by small figures in hunting scenes; and schematic art. And in terms of technological progress, the authors of Levantine art, from the Neolithic to the Metal Ages, were more advanced than the Magdalenians. They lived in villages and cultivated the land, they had grain, livestock, but they made art that was simpler in both technical and formal terms than the bison of Altamira. In reality, however, both kinds of art represent responses to different needs coming from different symbolic universes. The Bronze Age people could have made figures like the Magdalenians, but they didn't.

> The purpose of art and art itself have changed, but
> what was the purpose of the art that was made in
> caves 30,000 years ago?

The meaning of art is always the million-dollar question. When we walk into a cave and see these spectacular paintings from tens of thousands of years ago, we often wonder why they were created. But that's where the ground starts to get shaky. Today, we believe that Paleolithic art, like the art that came after it, was a form of language, a means of expressing ideas. The problem is that we don't have a point of reference now for what it meant. We recognize the signs, we can see them, we can study them, we can measure them, we can find out what they're made of, where they were made, and with what intent – or at least whether they were made to look a certain way. We have a tremendous analytical ability to describe them through the use of technology, but we don't have a Rosetta Stone of prehistoric art that lets us get any closer to deciphering their meaning. There have been many explanations since the phenomenon began to be studied in the last third of the 19th century: from art for art's sake – the idea that these representations responded to the aesthetic pleasure of making and seeing them – to art as an element of propitiatory magic. This last theory, which was widely accepted for a long time, was based on the fact that drawing animals in conjunction with projectiles, especially around the belly and flanks, was seen as a way of contributing to their capture. This interpretation was also supported by ethnographic analogies drawing on the knowledge of groups of hunter-gatherers who survived well into the 19th century.

> How is it interpreted from a more symbolic
> point of view?

There have also been totemic approaches to interpreting some of the geometric signs because rock art, aside from representing animals, is 50% an art of signs. Needless to say, the animals are also signs. When we see a horse, we know that it isn't a portrait of a horse, because rock artist weren't portrait painters, even if they were able to identify incredible anatomical details. Rather, the idea is that the horse represents something more than the horse, which relates to cosmogonic traditions or mythograms. That said, normally we use the term "sign" to refer to geometric patterns that seem to have no correlation in the natural world (compartmentalized rectangles, circles, and other geometric shapes). The totemic magic relates to the symbols found in some regions, which serve as identifiers of various ethnic groups. Later, around the 1950s, structuralist theories began to be adapted to interpret Paleolithic art as a symbolic representation of universal general principles, often contrasting two central principles associated with male and female poles. According to this view, a series of animals that are clearly paired together in the paintings are associated with the male and female principles. For example, bison and horses, or horses and reindeer, are structures that are repeated in Altamira and in the Cave of La Pasiega. According to this interpretation, the murals represented a general idea of the cosmos governed by these two great principles. Of course, that reading could be applied to some of the paintings but not all of them. Nevertheless, that perspective was predominant until the 1980s.

> What do contemporary interpretations of
> Paleolithic art say?

Since the 1980s, those overarching theories have fallen by the wayside to be replaced by more postmodernist interpretations, in which rock art is valued as a tool for interpreting Paleolithic society as a whole.

Lately, the number of interpretations has been atomized and a number of very interesting new theories have emerged, like the one I mentioned earlier, in which rock art is understood not only from the point of view of the artist but also of the spectators. That's a social interpretation of rock art, not so much in terms of the structure of the society that it reflected but rather the artistic activity itself – and the entire cave – understood as an installation. Now, in terms of meaning, I don't think there's one single explanation. Rock art as a phenomenon was in existence for at least 30,000 years and on five continents. Every continent except Antarctica, that we know of. It should be noted that it was practiced by hunter-gatherers and more sedentary societies, which, although they were different, had the same need to express some of their ideas on the walls. It's highly unlikely that such a broad phenomenon, associated with such different societies, could have one single explanation.

> Does this interpretation of art as a social event tie in
> with the idea of artistic practice as a mechanism for
> forging connections between individuals in a group,
> as a cultural tool to bring society together?

There are many interpretations for that as well. The more Anglo-Saxon view argues that culture is a system to help us adapt to our environment, in almost economic terms. The French, on the other hand, look at rock art as an eminently social art. Then there are anthropologists who interpret it as a personal, individual fact, which refers back to the more Romantic or old-fashioned idea of the artist. In contrast, there is an intriguing expression that calls Paleolithic art "an art without artists". Clearly, there were particular individuals who painted, engraved and sculpted. In fact, the roof of Altamira is very different from the rest of the cave. It's clear that they were done by different people. There are even times when a certain group can achieve what seems to us to be the pinnacle of artistic creation, but the interpretation of art as a social phenomenon prevails. Because Paleolithic art, despite being a naturalistic art in which artists drew the horses' eyelids in great detail or drew the hairs in certain areas with incredible precision, is also full of conventionalisms. That is to say, there are ways of drawing the figures and executing the paintings, specific ways, for example, of representing the various anatomical parts, that are repeated from Asturias all the way to the Pyrenees and the Dordogne. There are figures made from bones that you could interchange between sites situated a thousand kilometers apart, and no one would notice. Or paintings of bison on the walls in Asturias that are the same as others in the Pyrenees. That means that not only is there a shared sign that it is the same over a distance of thousands of kilometers, but the way the sign is represented is also shared. Those signs or figures are models, true prototypes. That means that there is an ideological and technical community that spreads over thousands of kilometers. There is a shared model that isn't tied to the creative genius of any one individual but to social norms and how those norms or ideas traveled at a time when the climate conditions were very harsh. We have to remember that these people travelled on foot; they moved slowly, but they weren't isolated. They shared common codes and forms of communication.

> Along those lines of common symbols, *Homage to the Elements* uses the spectra and the periodic table. Plus a lot of Balcells' pieces also incorporate a cosmogonic drive or a desire to represent the entire universe.
> Does that happen in Paleolithic art?

Along with the view of rock art as a social event, independent of singular brilliant or creative individuals, there are also theories that describe it as an art that explains the world. We believe that the figures on the walls are the remnants of a series of activities, which would also have included music. Flutes have been found that are very well preserved, but there must also have been drums made from leather skins, and probably certain string instruments because they knew how to use vegetable fibers and animal tendons. That's all that is left of those activities, those performances that took place inside the caves, during which people no doubt sang songs and told stories, activities that also help bind groups together. Lots of groups achieve a certain social cohesion by drawing on a common past, which may invoke ancestors with names and surnames or mythological beings who preceded humans, or even territories. Therefore, these paintings, as illustrations, were probably part of an activity that was meant to explain the world. Human beings have always asked ourselves questions that we can't answer. Now we have a technical and scientific development that moves forward based on research, debate and controversy, and that explains the world to us. But back then, when there wasn't all this knowledge, it's very likely that rock art was part of those attempts to explain the world and to explain human beings' place it. Who are we and why are we here? Where do we come from and where are we going? In short, the eternal questions of humanity, without necessarily implying a religion as we understand it today. Because a religion isn't just an individual belief but a structure that has participants and officiants, a hierarchy and a series of roles. If we look at it from that point of view, Paleolithic art would be more like a kind of religiosity with close ties to the natural world, like animism. In any case, the paintings probably have to do with a cosmogony that is meant to explain what the world is, where it comes from, and our place in it, our role. A role that also amazes us and terrifies us because night falls and then the day breaks, because there are variations with the changing seasons, and there are storms and lightning and seas and rivers... natural phenomena that still impress us today, and for which the people living in the last glacial period would no doubt have sought an explanation.

Move
Before Thinking

María Muñoz

I have to confess that before I talked with María Muñoz, I would only listen to Glenn Gould's rendition of Bach's *Goldberg Variations*. More specifically, his recording from 1981, which is more serene but more emotional than his 1955 version. It was just an obsession of mine, which I justify on the basis of a rather extensive search for versions by different performers – from Lang Lang's more romantic and embellished rendition to Rosalyn Tureck's more slow and delicate touch. I wasn't familiar with Dan Tepfer's version, in which he takes the liberty of not just playing the variations, like everyone else does, but of including long, jazzy and yet still baroque-at-heart improvisations that offer a reinterpretation of the variations that is very personal and, following the original spirit of the piece, playful and lighthearted. Variations of the variations that ultimately play along with the game Bach suggested. Before speaking to María Muñoz, it had never occurred to me that you could dance to the *Goldberg Variations*. But she does just that. And she does it by following a process that includes a lot of interpretation, and a lot of research as well. So what is a dancer looking for in the *Goldberg Variations*? She doesn't know until she starts looking. And what does she find? Well, something that obviously can't be described in words, but which has captivated me with a certain amount of fervor. After talking to her and reading a little about Tepfer, I decided that I wouldn't stop until I could watch them together – him playing and her dancing the Variations. And to top it off, Tepfer hums as he plays, just like Glenn Gould. I have no more excuses. Gould would understand.

What is your overall impression of *Homage to the Elements?*

Eugènia Balcells' approach to this piece, like in *Frequencies* and *Universe*, which together form a kind of very profound trilogy, has a lot in common with dance. I see the driving force of an enormous curiosity in it, that puts whatever interests her at the center of her research. And that, understood as a creative process, is very similar to how we approach creation in dance. We start from a void, a darkness, and from there we try to approach what interests us through intuition. And of all the tools we have at our disposal, the most important for me is the ability to make connections, to associate ideas that come from everything you see and how discover things. That association starts the engine that drives you forward. And as far as the content goes, Balcells' proposal is very powerful, because it looks at the matter that makes up the world and aims to discover what it is and how it is manifested. That's an immense topic that I approach from a more fragile perspective, based on the human body. *Frequencies* is a dance of spectra that every now and then pauses, creating moments of fixation or silence in which an element appears. For me, that's also movement and dance. You have your whole body, open and sensitive to your surroundings and you let the movement emerge, often without wondering why. And then, from time to time, there are these moments of suspension where everything becomes clear.

Homage to the Elements starts with the elements, which are a material entity, but it focuses on the light that each one emits, which is unique and specific. We all have a body, which is matter. How we use it is more ethereal and is also unique to each of us. In that sense, could dance be considered a kind of signature or a deeper manifestation of the body?

We often use the idea of presence in dance as a concept related to the dancer's entire body, technique, and intuition. When you're dancing, you prepare all that so that something inside you can be expressed and released. We call it presence, but there are other names for it. Sometimes we say that a dancer has a certain light. And we say that about a body even before it starts dancing, as soon as it takes the stage. In that sense, there's a very strong connection with Balcells' piece:

each element emits its own light, and you could say that the same thing happens in dance. But what is that interior of ours made of? It's a mystery. But it's clear that memory, with all its aspects related to invention, is a part of it. It's also made up of one's own experiences, that ones you control and can explain, and ones you don't control but which are recorded in your body. There's also the ability to imagine, tell stories, and create mental images that push you into movement. That interior is made up of many elements and, ultimately, you work with your body to bring all that out. Sometimes the work is done in a mechanical way that involves thinking about all your tendons, muscles, and organs. Maybe not all dancers do this, but for me, it's important to think about that interior place and whether or not my organs are letting it out. And then, obviously, the ability to bring whatever comes out into dialogue with the exterior world comes into play, with all its implications in terms of proportions, perspective, projection, and generosity.

> In *Frequencies*, the light from the moving spectra forms a creative process. The movement allows the elements to combine and give rise to new matter. Ultimately, movement – and dance – can be understood as a metaphor for creation: if we're holding still and not expressing ourselves, nothing happens, nothing is created.

To get somewhere and find textures, as we call it, or ways of being, you need to get moving. And, often, you have to not overthink it. On the contrary, when you give the movement a premeditated intention that you want to express, that's when nothing happens, because you have put a screen in front of you, a controlling mind that wants to decide how the meaning of what you're doing should be organized. In our work, we've always kept that in mind, and we've reasserted it. There's something much more basic that is related to the world, to the movement of the world, and to the transformation of things, and also to energy: get moving, get your body going, with its limbs and their articulatory and associative abilities, and it will give you the starting point for investigating something. And often the final expression, or how it enters into meaning in the context of our culture, comes later. That's why dance has something very preverbal about it that makes it possible for people from very different cultures to communicate.

However, if it came down to interpreting what you're doing, everyone might use different words.

Is this preverbal nature also present in Balcells' work?
She is constantly working on ideas that escape definition and labels and boxes. That's why she often talks about poetry because if poetry has something to offer, it's that you can't tame it or censor it because you never know where it may appear to someone. And I like that. Although the themes may sometimes be very technical and often involve mathematical formulas, the way science approaches them and how the paths have found to discover that knowledge is very similar to what happens in creative processes. And I also think that in science, when everything starts to be classified and closed off, the research peters out. It's precisely when there is a moment of intuition or a more or less crazy hypothesis when everything is set in motion. I really like that about *Frequencies.* I spent many hours in front of the installation. The movement of the lights and colors has a strong impact on our imagination, and you get totally hooked. It's mesmerizing. Then you get the explanation that every light comes from an element, but while you're there, you're just perceiving it. There's something very alive about the piece.

Movement and dance, like the interactions between elements, are a process of transformation in which something invisible is made visible – in this case transforming the space and the perception of the spectator.
Dance can be understood as a way of finding your own voice, your body's voice. And in our whole process, it's clear there is something transformative. Even though we may be small, imperfect, and very limited, the body is endowed with this spirit of movement. It makes what you have inside vibrate, sets it in motion, and has the ability to release energy – an energy that, when transformed, can sometimes come out from any part of your body. For example, a few years ago we were doing research on Baroque counterpoint, and I wondered if it would be possible to bring multiple voices into a single body – as though one arm had one voice and the other had a different one. This was related to the idea of the canon and the fugue in order to understand the body as a harmony. A harmony in which, through the breath, the mind connects with the center and, from there, builds a

house for its presence, which in the end is a simple vibration that takes you from being in the present moment to generating this energy. The body, as matter, has that voice in movement.

> For dancers, the body is your means of expression, but it is a medium that changes over time: a 20-year-old body is not the same as a 60-year-old body. Can a changing body transmit the same light, the same emotion?

There is a change, it's true, but it's for the better. When you start out, there is such a strong desire to have what you can only achieve after a long journey of experience that everything moves in a chaotic way, as if the elements were constantly colliding with each other. You want to talk about tenderness and your body is tense and you make movements that you can't control. In the beginning, there are a lot of paradoxes. There's a strong desire, but there isn't the serenity of understanding that some things take time. Also, apart from experience – and I think this can be extrapolated to other areas – from a physical point of view, as you grow there are changes, there are more limitations, there are things you can't do anymore or don't want to do. But on the other hand, you realize that experience, the journey, the process is powerful in itself. Experience transforms you and allows you to be more exposed and vulnerable. And in general, in our society, vulnerability can be found on the periphery. We tend to focus on what is the strongest and the most enterprising as though that is the center of the path, and we often fail to look into the far corners. So then, when you get to the point where your vision broadens and you pick up the edges, you realize that there are small spaces of light in that vulnerability. That light might not have as much projection as the kind that comes from strength or muscle, but deep down it is a stronger light because it comes from a deeper, more interior place.

> For the performance *All the Names* you researched identity, how classifying and naming everything determines it and pigeonholes it, and the fact that sometimes changing those names brings about a very important reevaluation. *Homage to the Elements* is based on one of the most extreme and objective classifications, the periodic table. What process did you follow to research this act of classifying or naming?

In creative processes, you sometimes have a clear starting point and other times not so much. In this case, I had been thinking about the topic of naming for a long time, about how something is comes into the light when you name it. Then I found a book by the philosopher Carlos Thiebaut, *A History of Naming*, in which he lays out a real or hypothetical history of the ways in which we construct our identity. After a theoretical and methodological introduction that questions the relationship between the concepts of name, text, and identity, Thiebaut proposes a much more personal and creative second part. It fascinated me how, from his adult memories of seeing two paintings in the Prado when he was a child (*Jacob's Dream* by Josep de Ribera and *The Drowning Dog* by Goya), he describes the idea of two different ways of naming: the ancient way, which involves receiving a religious and political identity; and the modern way, in which our identity is what we give ourselves, a complex construction of subjectivity. And I contacted him. When we started exchanging ideas, it was clear he was open, creative, and willing to share in the adventure of creating *All the Names*. So, I took those ideas into the studio to improvise.

And what came out of it?
A process in which the ideas were completely transformed, to the point where Thiebaut said, ironically, that his doctoral dissertation was useless. We invented a character, Carnival, who introduced himself by defending the freedom to be able to change his name, but at the same time revealed a confusion because it's unclear where what he's saying is coming from. There's a paradox in both the research and the final texts of the show, which is reflected in the fact that Carnival says one thing and, later, contradicts himself. In the end, these two faces, the beautiful part and the demonic part of giving and receiving a name, gave rise to a lot of ideas and material for a very emotional performance. We also shared the process with John Berger, who wrote the words that concluded the show:

"Without a name, we are naked.
Being naked is a form of waiting.
You are waiting for them to give you a name. The first one."

Berger and I reflected and worked a lot on the topic of identity and for another performance, he gave us this text:

> "Tell me, where does a body end? Outlines delimit things, are useful for maps and sometimes for drawings. But do they really exist? Do bodies have borders? And assuming they do not exist, what about this thing we call identity? Perhaps identity is nothing more than a temporary agreement between several things, a curious type of complicity between separate elements that would otherwise have nothing to do with each other. You say we are everything, and if we are everything, maybe identity no longer matters. Go and ask the shepherds in the mountains about identity."

> You also did a piece inspired by Bach's *Goldberg Variations*, which are variations on the same theme – like Balcells' work shows that, based on a few elements, combinations can be created to give rise to a vast variety of matter.

In that case, we looked at whether it was possible to transpose certain compositional tools used in Baroque counterpoint onto the work we do moving our bodies. Bach is the master of showing how to create structures of great complexity using a few notes. During the process, we worked on the recordings made by Dan Tepfer, a pianist who, in addition to playing Bach, is a great jazz improviser. He made his own variation from each Variation, with the same harmonies and respecting the wording and structure. So, in the end, we did our own variations of the Variations that are the result of a process like the one we discussed at the beginning in reference to Balcells' work, a process based on working with the body and this idea of setting a limited number of elements into motion to start down the path of creation.

The Art of Looking at the Sky

Marc Balcells

I am convinced that everyone, if even unconsciously or unadmittedly, has a weakness for astronomical observatories. These concrete and steel structures the size of a building that weigh dozens of tons and move with a precision to the nearest thousandth of a millimeter can't possibly leave anyone indifferent. Moreover, these true monuments to human ingenuity and knowledge are often built in remote places. On mountaintops, in the world's most arid deserts, or on the edge of a volcano's cone. The reason is simple: there is less light pollution in those places, the air is thinner and drier. That means the light coming from stars, nebulae, and other celestial objects is less affected by the atmosphere. The telescopes in these modern observatories are no longer tubes you put your eye up to. They are sensors that transform light into data, which is sent to a computer for processing. Modern astronomers don't look through telescopes anymore, they use computers. But even so, they still have a certain romantic way about them. There's a touch of enchanting fatalism in facing the absolute blackness with an inquiring mind and the help of a machine, knowing from the outset that we may never understand the innermost nature of the universe. Because every step of this intellectual adventure that consists of unraveling the mechanisms of nature is a worthwhile step. Marc Balcells is that kind of astronomer. You can see it in the way his eyes light up when he explains how scientists discovered that the elements were formed in the cores of stars in tremendous reactions of nuclear fusion. Those nuclear processes, governed by laws that we have been able to reveal, are responsible for the existence of chemical elements, for the fact that it was possible to organize them on the periodic table, and for the possibility of works of art being created based on that table. We are, therefore, literally children of the stars. And, as Marc Balcells points out, we know this thanks to the spectra that served as the foundation for the creation of a work of art like *Homage to the Elements*.

You and Eugenia Balcells have known each other for
a long time.

She's my older cousin, and we've been in close contact since we were little. She would come to spend the summers at a farmhouse near Girona, and we would see each other and talk a lot. For us, having a cousin who had been to the United States was amazing. We were all very musical at our house, and she really liked our family. When she got married, she asked me and my brothers, four little kids between the ages of eight and twelve, to play at the wedding, and we did. We had a real bond. Later, when I went to do my PhD in the US and she was living in New York, I visited her a few times. And during these meetings we talked a lot about the relationship between art and science, and about concepts like space and time, both with her and with the artist Peter Van Riper. Since I was an astronomer, they were very interested in my point of view.

Have you followed her artistic career?

Absolutely. Every time she opened a new exhibition I would go, and we'd get together and talk. Later, we were farther apart geographically because I was living in Holland first and then in the Canary Islands, so we didn't see each other as often. But in any case, we kept in touch and continued talking about science and, above all, about the idea of space and mankind's position in that space, about humanity's place in the cosmos. In fact, the cosmos is Eugenia's ultimate focus. She has always had this idea of uncovering the place from which a person views their surroundings. In the late 1970s, she created *From the Center*. From a fixed point, the roof of her building in Manhattan, she filmed everything going on around her, taking the four cardinal points as a reference. This piece is a metaphor for what astronomers do with respect to the cosmos: humans are here, standing on the roof of our house, looking up at the sky in all directions, using telescopes instead of cameras.

At one point, Balcells' interest in science took came to
life in *Frequencies, Homage to the Elements*, and *Universe*.

Those pieces make up the exhibition *Light Years*, which is a visual and auditory metaphor for space and the cycles of time. She also put the periodic table under a prism to provide a kaleidoscopic view of the universe. I remember now that one summer, when we were little, Eugenia brought a kaleidoscope and we looked at it with intense interest, discussing how it worked. *Frequencies*, *Homage to the Elements* and *Universe* are different prisms for approaching reality. I would say that the years in America helped her to realize, among many other things, that science offered a unique way of looking at the world, a vision she needed in order to work more profoundly with the ideas of space and time. Plus, in *Universe*, there is an element that is very characteristic of Eugenia, which has to do with describing, in a very original way, the fact that there is a cyclical part to the universe: day, night, the moon, months, years, the orbits around the sun, and the final "year" or revolution that exists, the 70 million years it takes for the sun to travel around the galaxy. But from that point, the principle changes and things don't come back around again. Everything that happens only happens once, and that's it. Every moment is unique and unrepeatable.

> What did you think of *Homage to the Elements* when
> you first saw it?

I thought it was an impressive synthesis of previous pieces. In 1976, she had created *Supermercart*, one of the first pieces she did when she returned from America. In that instance, she put all sorts of objects into plastic bags. There were leaves, scraps of paper, etc. The piece incorporated the idea of taking reality and putting it into a box to be able to look at it. The periodic table, in the end, is also based on taking each element and putting it in a box. But since you can't really put iron dust or a little bit of helium on the wall, she had the brilliant idea of using the emission spectrum for each element.

> Both *Frequencies* and *Homage to the Elements* are
> based on the spectra of light emitted by each
> element. How important has the concept of a
> spectrum been in the history of science?

It has had a tremendous importance. The periodic table itself was already an incredible discovery for humanity. It should be on display in every public square in every major city, and Gothic cathedrals should have a chapel dedicated to the periodic table in the place of the saints and martyrs. As for the spectra, they help us to learn things about the elements. Atoms emit light with certain specific frequencies as opposed to others when we excite them. When someone touches an atom, it vibrates and reveals its internal resonance. And just knowing that atoms aren't sealed off like a billiard ball but that they respond like a butterfly, is already amazing. In fact, they're like us; if you prick us we bleed, like Shakespeare wrote. In that sense, atoms are alive. And the spectra made it possible for us to discover this intimate aspect of nature. Plus, because it's a vibration, the fact that every atom emits a characteristic wavelength has a musical connotation. Each type of atom has its own resonance. Some of them have high voices and others have deeper voices. Another interesting thing is that scientists understand these spectra that make up Eugènia's periodic table in the tiniest detail. We can write an equation that tells us exactly what their frequencies are, and that's an extraordinary milestone in knowledge. Because it means that we really understand the laws of nature. From a more astronomical point of view, thanks to the spectra we know about the makeup of the cosmos, and, moreover, we know how much of each element there is. In other words, we know if there is more hydrogen than helium, iron, or oxygen. And it also lets us know the temperature of the things we're observing and their density. We know that out there in space, there is approximately one atom per cubic centimeter – in other words, about a thousand atoms per liter. It isn't a lot, but we can measure it. And yet, here, around us, in this room, there are 10^{23} atoms per liter. Spectra can also tell us how fast things are moving and whether they are approaching or moving away. The expansion of the universe was discovered because the lines of the spectra weren't where they should have been; they were shifted toward the red end of the spectrum. And the farther away the galaxy was, the more redshifted its spectrum was. That gave rise, for the first time, to the idea that things were moving away, and then came the conceptual shift: it isn't that the objects are moving away, but rather space that is expanding. We're laying bare the universe. It isn't hidden anymore. We've pulled back the veil, and we can see everything because of the spectra.

You've said that "if a picture is worth a thousand words, a spectrum is worth a thousand pictures".

I didn't come up with that myself. It comes from a rivalry that is ongoing between astronomers. There's one cohort that works with spectroscopy – i.e., the study of spectra – and another that captures images of the sky. The people who rely on spectroscopy love that saying, but the amount of information that can be extracted from a spectrum is really huge.

The existence of spectra was discovered even before people understood why they existed. In fact, helium was discovered in the sun before it was found on Earth.

That supports an idea that runs somewhat contrary to what we learned in school: that knowledge is empirical. In school, they tend to teach you the theory and then you apply it. There are still astronomers who are more theoretical and who want to begin with a theory and then prove or refute it, but all the progress that has been made in astronomy has been thanks to the empirical act of looking, because we've found things we never could have imagined. Newton had already discovered that when sunlight passed through a prism, the colors separated. And then he added another prism, and he saw how they recombined. In the 19th century, when sunlight began to be observed in more detail, they saw a series of lines that didn't correspond to any known substance on Earth, and in reference to the Greek name for the sun, they called it helium.

How does a scientist look at works like *Homage to the Elements*?

When you see the periodic table somewhere, the first thing you think as a scientist is, "It's great that someone has taken an interest in this." And when you find that it sparks a big response, you're even more amazed. I have to confess that the first thing I did was check that the spectra were correct. Another thing I saw right away, as a scientist, is that the periodic table is too anthropocentric. Because astronomers don't just know what elements are in the universe, we've figured out how they were created. Immediately after the Big Bang, there was just hydrogen, helium and a little lithium. Then the first stars formed. A pair of English astronomers, Margaret and Geoffrey Burbidge, wrote an article about this in 1957, which led to a giant leap in our thinking about the elements. They proposed not only that stars get their energy from converting hydrogen into helium, but that the helium is later converted into carbon, and carbon into oxygen, in a process that eventually results in all the elements on the periodic table. Therefore, all the elements were formed in the stars. And that has become one of the most revealing truths in our knowledge of the cosmos. It was one of those times when just a few people realized something, and it turned into an advance that transformed knowledge for all of humanity. Since we know a lot about this process now, I'd like to turn the periodic table upside down, with hydrogen and helium at the bottom, to see how the rest of the elements are formed, like a building being built starting with the structure.

> In astronomy, looking far away means looking at the past.

Astronomers don't just see what elements are present, in what quantities, and how they move. Because light takes longer to arrive from distant objects, when we look at them we're seeing them in the past. The farther away they are, they farther back in time we're looking. Using powerful telescopes, we can put together a portrait of the past universe, reconstruct its history, and gain a perspective view on the evolution of the cosmos. We've seen stars from 11 to 12 billion years ago.

> Can works like *Homage to the Elements* contribute in some way to generating interest in science or making it more accessible to society and lending prestige to scientific culture?

I think, above all, they can have a major impact on young people, as sensory pieces that affect you without the need for much of a conceptual element. I'm thinking of *Universe*, for example: when you walk into that dark room and you see everything spinning, you're looking at things you've seen before, like the moon, the planets, or the galaxies, but you're seeing them in a different way, with a movement that really touches you.

A Table
Full of Stories

Joaquim Sales

The periodic table is a cultural icon. It has been turned into shower curtains, coffee cups, T-shirts, provocative underwear, neckties, and a long list of other merchandise. And that isn't surprising, since it's one of the scientific achievements that most faithfully embodies the desire of science to explain as much as possible with minimum means. In that sense, the periodic table contains everything. Everything we know about, that is. All the ingredients and all the possibilities. In other words, all the stories. As Primo Levi said in his marvelous and fascinating book *The Periodic Table*, there are so many atoms that you can always find one whose story coincides with one that has been invented at random: a molecule of two nitrogen atoms in the air of a concert hall that vibrates at the frequency emitted by a violin playing Vivaldi; a silver atom stuck to the nitrate film the Lumière brothers used to film the workers exiting their factory; or a carbon atom in the ink used to print this period: this one here. Joaquim Sales knows these stories and many more: the stories that tell how the now iconic periodic table came about – a process that is still ongoing as a handful of laboratories around the world prepare new elements for the first time. And the stories about how knowledge develops, in addition to being stories, are also knowledge in themselves, because they reveal the winding paths and changes in direction that the human intellect follows, without exception, in any creative process. It is a complex journey that reaches its maximum explanatory potential when it is put into the form of a series of words one after the other – that is, once it is translated into a story. And Joaquim Sales has been in possession many of those stories for years. Stories which, like the xenon atoms in a camera's flash tube, also illuminate the world.

As a chemist who is an expert in the periodic table,
how did *Homage to the Elements* strike you?

I saw *Frequencies* first, the installation where the elements' spectra combine
with one another. I thought it was stunning. As a chemist, I was aware of
the spectra of the elements and the periodic table, but that representation
somehow made the uniqueness of each of the chemical elements explicit in
a very original way, and that was incredibly powerful. I remember that there
was a periodic table as part of the exhibition, a kind of rough draft of the
mural we now know as *Homage to the Elements.* When I and other chemists
saw it, we thought it would be worthwhile to delve deeper into that idea.
We suggested it to Eugènia Balcells and she did it. She added a new, very
special periodic table to all the others that already exist.

Are there a lot of them?

Hundreds or even a few thousand. There was a boom in 2019 for the
International Year of the Periodic Table. People adopted the periodic
table model to talk about all kinds of things: birds, jam, chocolate,
typography. As a representation of the chemical elements, there are also
many different versions. Funnily enough, when you see this one, you
think: "How is it that no one had done this done before?" The most
amazing thing is that it was done by an artist! It is a powerful, compelling
and enthralling contribution. We bought it for the chemistry department
at the University of Barcelona, and today it presides over the reading
room in our library.

As for all the other periodic tables... do people use them?

We have a collection of them in the chemistry department. Some of them
are historic, and others include the properties of the elements. We have
them hanging in the hallway, like many laboratories do, but the one we
use most is the traditional version, the one that's printed in most books.
Because it clearly shows the behaviors of the elements. The first thing
you see right away, for example, is the line between the metallic and non-
metallic elements.

Are chemical elements often represented in art?

There are a lot of representations, but I remember one in particular
that is not very well known; I came across it as a tourist at the National
Gallery in London. It represents the elements as they were understood
by the ancients: fire, earth, water and air. It's a series of 16th-century

Flemish paintings by Joachim Beuckelaer. One painting is of a bird, which represents the air; another is of fish, which represents water. There is another that shows game about to be cooked, which represents fire; and another full of vegetables, which represents the earth. It's a symbolic representation. But *Homage to the Elements* uses the elements themselves, represented by their emission spectra, and everyone can see them and incorporate them in their own way. In that sense, having had that vision is a great accomplishment for Balcells.

As for the relationship between art and science, as manifested in *Homage to the Elements*, there are research centers today that host artists in residence, who spend weeks or months in the company of scientists in order to create a work of art. They do it at the European Organization for Nuclear Research (CERN), for example, and in Barcelona there are centers like the Biomedical Research Institute that do the same thing. What can an artist's perspective contribute to the scientific world?

I think that art and science are two very different worlds. There's no doubt, however, that artists can draw inspiration from the world of science. Residencies can help artists understand what it's like to work in a lab, and maybe see something there that only an artist would see.

Where did the need for the periodic table come from?

The idea of an elemental substance came from the Greeks. They had already conceptualized the four elements. Then, in the Middle Ages, the alchemists tried to make gold, because they had the idea that it was different from the other substances. But the first definitions of the elements came from Robert Boyle in the 17th century. He argued that they were substances that couldn't be broken down into simpler ones. Later on, Lavoisier (1743-1789), who is considered the father of chemistry, made an initial list of the elements; there were around 50. And when Diderot and d'Alembert created their *Encyclopédie*, they realized that the elements needed to be organized in some way.

It wasn't enough just to list them?

People began looking for analogies between the elements with the most similarities and developing different proposals for their classification. John

Dalton (1766-1844), who devised atomic theory, laid the groundwork for the definition of chemical elements with the idea that each element is characterized by an atomic weight. In fact, in his book from 1808, he laid out the first table of elements organized according to their atomic weight. From there, other people began making other attempts at classifying them. It was a long process, with many different contributions. An important moment was in 1816, when André-Marie Ampère offered a classification by chemical properties. Then, in 1820, Johann Döbereiner began grouping elements into triads, because he had seen that there were groups of three where the middle element had properties that were midway between the other two. Another significant point was in 1862, when Alexandre-Émile Béguyer de Chancourtois proposed a spiral arrangement for the elements: the so-called *vis tellurique* or telluric helix. He put hydrogen first and then he realized that the properties were repeated every certain number of elements. He began intuiting the periodicity of the properties as a function of atomic weight. At almost the same time, John Alexander Newlands (1837-1898) organized some 60 elements by atomic weight and observed that a given element has properties similar to the eighth element that follows it on the table. From there, he introduced his famous law of octaves: "The members of the same group stand to each other in the same relation as the extremities of one or more octaves in music." And then came Mendeleev.

He completed the process.
Exactly. Dmitri Mendeleev (1834-1907) is the father of the periodic table. Although there was a bit of a controversy a while back because a German chemist, Lothar Meyer (1830-1895), developed a similar table around the same time. But hardly anyone remembers Lothar Meyer today.

Why is that?
Because Mendeleev's main contribution – the big difference with respect to Meyer's proposal – was that he left empty spaces on his periodic table. He predicted that there were elements that hadn't been discovered yet, which would be found in the future. And that turned out to be true.

Amazing.
He even predicted the properties of some of the undiscovered elements, like their color or density! And all that has ultimately meant that the periodic table is Mendeleev's periodic table.

He didn't get anything wrong?

One thing that got under his skin were the noble gases. He had made the table before they were discovered, and of course they didn't fit. He even went so far as to claim they didn't exist. He was stubborn, but ultimately he accepted them and worked out a solution: he put them in a column at the end of the table so that everything lined up. In other words, in the end, his fear or reticence toward the noble gases actually served to consolidate the periodic table.

Being able to predict not only the existence of unknown elements but also their properties is a demonstration of science's capacity for a profound understanding of the world.

You're right. And that's what science is. I think the periodic table is one of science's greatest contributions. I remember when I was at school, I had a math teacher who was quite a character; for him, the world could be divided into math and trivialities. And he said that one of the few non-trivial things outside of math was "that little table."

Was Mendeleev's original periodic table very different from the one we use today?

The first one he designed in 1869 ran in the opposite direction. Then he changed the rows into columns and turned it, but the principle was ultimately the same. Mendeleev's table had eight columns and now there are more, which correspond to the transition elements. And there are two additional periods, situated at the bottom, for the elements called lanthanides and actinides. That's why it's called the long form, but it's essentially Mendeleev's.

There are a lot of people who imagine the periodic table as something that hasn't changed in 150 years, but elements have continually been added.

And still more will be added. We're currently at element 118, oganesson, which is a noble gas and marks the end of the seventh period. In Mendeleev's time, the table ended at uranium, element number 92, which is the last element found on Earth. In the 1940s, and coinciding with experiments on the atomic bomb, scientists in California began preparing transuranium elements. They were called transuranium because they came after uranium. Eight more were prepared, through lawrencium

(Lr), which completed the actinide period. In the mid-1970s, there was an impressive step forward with the preparation of so-called superheavy elements, which completed the seventh period.

That happens in laboratories, right?

Right. In highly specialized laboratories. There are only three or four of them in the world: in the United States, in Russia and in Germany. Essentially, they run experiments in nuclear physics. Atoms from light elements are used in bombardment, and new elements are created through processes of fusion. In fact, only a few atoms of each new element are formed.

These atoms are made artificially, and they last for just a millionth of a second. Can they still be considered elements?

They are elements, even if only for a very short time. Once they have been prepared, analyzed and characterized, they are elements. For now, of course, they have no direct applications.

The physicist Richard Feynman said that if all knowledge were destroyed and only one idea could survive, he would choose the atom. Understanding that everything is made up of atoms offers an understanding of a very fundamental aspect of nature. And the periodic table communicates that.

Exactly. Everything material in the universe is there. There are only combinations of those elements. The Greeks actually said the same thing: there is the atom and there is nothingness, a vacuum.

Why is the periodic table called "periodic"?

When you classify the elements, their properties repeat. Looking at any column from the table, you'll find similarities in the chemical properties of the elements. That's where the table got its name.

Why are there similarities between the elements?

Because they have a similar internal structure (called the atomic structure). The distribution of electrons in layers that characterizes chemical properties is similar for elements in the same group or column. Because, essentially, chemistry is the exchange or sharing of electrons. When two elements share and/or exchange electrons, a chemical reaction occurs.

From there, we can predict and interpret the chemical behavior of the elements and compounds.

How are the names of the elements chosen? The last ones are a little strange, aren't they?

How the elements get their names is an interesting topic. When a new element is prepared, the scientists involved in the process have the right to name it. There is even a nomenclature. During the preparation and throughout the acceptance process, the IUPAC (International Union of Pure and Applied Chemistry) has a set of rules for giving an element a provisional name. Once the discovery has been recognized, the scientists can give it a definitive name. The names usually refer to people or to the geographical locations where the laboratories are located. Dubnium, for example, was named for the Russian city of Dubna, where it was discovered.

The process seems pretty straightforward.

Not always! In the late 1990s there was a very interesting, very heated controversy between the IUPAC and the American Chemical Society. The IUPAC had established the criterion, which I personally agree with, of not naming elements after people while they are alive. It makes sense to me, even in the case of something more commonplace, like street names. This led to a very heated debate between the two organizations because the Americans wanted to name an element seaborgium, in honor of the American chemist Glenn Seaborg (1912-1999), who was one of the fathers of the transuranium elements, and whose scientific talent was unquestionable. But he lived for a really long time.

So they couldn't use his name.

A series of committees were formed and lot of different proposals were presented, but in the end the Americans won out. Mr. Seaborg got his element, seaborgium, while he was still alive. The case was published in scientific literature, because it was basically a power struggle (*Unsuspected?*). Then, there were the French, who wanted to name it Joliotium in honor of Frédéric Joliot, who, together with his wife Irène Joliot-Curie (daughter of Marie Curie), won the Nobel Prize in 1935 for the preparation of new radioactive elements. Apparently, his communist past was part of the reason his name was eventually dropped from the proposal.

The Most Profound Expression

Rodolfo Häsler

The first time I read about the idea of self-formulation in literature, I had a revelation. In a way, I had self-formulated self-formulation. Let me explain. One of the things that literature does, perhaps the most important thing, is to put the human experience into words. There are times, when we've experienced a loss, a frustration, a passion, but all those experiences are clouded like a fog inside us. They're there. We have no doubt that we've experienced them, they seep into our memories and even impact the decisions we make. But we don't know much of anything about their nature. We don't entirely understand them. Perhaps it's because we haven't reflected on them enough, or because we haven't been able to think about them in the right way. And then we read a book, and we see some of those same experiences formulated in clear, vivid language. The text expresses exactly what was in us but that we weren't able to explain to ourselves. And a light comes on. I had experienced this kind of inner revelation as a reader before coming across the term "self-formulation", hence the play on words above. Reading Rodolfo Häsler's poetry, it happened to me again. During the lockdown, I think more or less everyone experienced what the Germans call *Fernweh*, which is essentially a certain nostalgia for faraway places, a kind of melancholy that comes from being unable to travel, unable to discover new worlds. In some of Rodolfo Häsler's poems I discovered a clear and vibrant formulation of that nostalgia: a feeling that isn't associated so much with the desire to do something that has become inaccessible to us for a time, but rather a defining drive – the vital need, as Fernando Pessoa said, "To travel! To change countries!/ To be forever someone else." If it weren't for that drive, would we have left Africa tens of thousands of years ago to eventually become who we are now?

> What are your thoughts about *Homage to the Elements?*

The pieces touches directly on the origins of existence, the mystery of us. The elements that make life possible are in us all. We are stardust, and Eugènia Balcells takes that evidence, which is often latent, and sets it out before our eyes. Plus, I think her work is deeply poetic. Poetry is one of the pillars of her work. She knows how to see what's there, so present and so astonishing, but which we've overlooked because of the distractions of everyday life. In that sense, her work is cathartic. It makes us pause, turn inward toward ourselves and awaken possibilities that make us better.

> The periodic table can be understood as a kind of dictionary of elements, which we might compare with words that combine to form the whole material world, which would be the equivalent of poems or novels.

I think so, the comparison is clear. For example, the combinatory logics that were used in music, poetry, painting and other areas in the 20th century show that there is an infinite space in front of us that is akin to mystery; they let us intuit that mystery.

> That reminds me of Raymond Queneau's book *A Hundred Thousand Billion Poems*, in which each of the 14 verses in each of the 10 sonnets that make up the book can be combined with the verses from the other poems, resulting in 100,000,000,000,000 different sonnets. If it takes a minute to read each poem, the book contains a million centuries of reading. Aside from the poetry, I've always seen a certain artistic intent in the book that transcends the texts themselves. In other words, there is a certain dialogue between poetry and conceptual art. From a more general perspective, how do you think art and poetry are related?

Every artistic expression is related, much more than society realizes. There are really no dividing lines, and there are as many points of intersection and communication as we can imagine. If there is no poetry in a creative activity, it has no depth or reason for being. We might say that poetry comes first. It's a prerequisite; without it, nothing can be true. In that sense, poetry accompanies every creation. It's that point of friction we pass through as we approach a work of art and something inside us shifts or is transformed.

> In your case, there is a clear relationship between an artistic discipline like painting and the poetry that you write. What is painting's influence on your poetry?

I always like being asked that question. Painting was my aesthetic education As you know, my father Rudolf Häsler, was a painter. Color is one of the first things that can attract a child. My father always had his studio at home – first in Havana, and then here – and I always experienced that as a gift. For me, coming home from school and seeing him painting was like walking into an endless, magical universe. During those years in Cuba, his paintings were huge surfaces of color, and he often spread the pigment across them with a spatula. Watching that colored paste spread, with its pores, its wrinkles and its different thicknesses, unleashed a sensual conception in me that took over my senses. As a child, I wanted to be a painter. As a teenager, that need for expression led to poetry, but that undercurrent is still there. It fuels me, saves me. It's a haven of goodness that offers profound satisfaction. Plus, poetry and painting have always had a productive relationship. They blend well together and pull each other in. Often, one helps explain the other. There are very interesting collaborations between painters and poets, and beautiful books where both pursuits communicate and challenge one another to grow. In my writing, painting is a constant presence. I might almost say it's an obsession – whether it's in childhood memories of my father's studio, in the presence of color, which offers so much abundance in giving names to things, or in looking at certain works, and even in a particular way of looking at reality, which is very pictorial.

> What does poetry consist of for you?

Some people have said that I write because I grew up in an environment that was conducive to creativity, and it is true that having that privilege may have been definitive, but I think that poetry, for me, is a way of organizing and understanding a very characteristic life experience: my parents came from two different countries, with two different languages and cultures. I don't live in the place where I was born or in either of my parent's home countries. I don't speak the same language now as I did before I was ten. In Barcelona I studied in German, which is not my native language... All those circumstances forced me to stop, observe, and discover. Occasionally, I've talked about how, when I arrived in Spain in 1968 – a country that was still underdeveloped and hermetic at that time – because my accent was different and I used language in a different way, I realized the weight that it carries. From there, writing is the next step. My writing is always based on a suggestion or a recurring memory that becomes something else. And recently I've been working a lot with obsessions that are part of my personal imaginary, such as certain cities, places, cafés, paintings, books. Poetry is inseparable from my everyday life. I'm not always writing, of course, but my way of feeling and looking at the world around me is entirely conditioned by poetry. I know that when I direct my gaze and I look at my surroundings, I can always discover a detail, a gesture, a phrase that I hear, a beam of light, an angle, that comforts me, that enriches me. It's a way of life that isn't cut off from everyday mundane things. It's about being present with a refined intensity. And looking a little further is often a blessing, it's true, but other times it can be painful because of the awareness it carries with it.

> Poetry, however, like Balcells' field of the visual arts,
> is more marginal than narrative or cinema.

Poetry, like any art that responds to an intense demand and a need, will always be in the minority, since it requires an effort of approximation on the part of the reader or the spectator. Poetry, and true art, is born from pain and from an interior wound. It is restorative and it can never come from a mundane attitude, so it will never be commercial. It is out of reach of the phenomenon of commercialization. Poetry can be free to some extent because no one is expecting it. The poet simply presages it and plays with it.

In Eugènia Balcells' work, I see a strong connection with what we were just talking about. From the very beginning, her work has been associated with using surprise as a way of accessing knowledge, with intuition – that leap in the spirit that disorients you and leaves behind a remnant of curiosity that pushes you to keep searching.

> Do you think that we should work to make poetry, and art in general, a majority pursuit and more accessible to people?

Our educational system would need to be radically reformed to give the humanities a predominant role. If we fail to feed our souls we'll always be crippled as human beings, without resources to draw on. Cultivating that sensitivity could help many people find an answer in the arts, and a spiritual pleasure, and a connection to beauty. If our education could teach that path, more people would be aware that poetry and the arts are as necessary as the air we breathe. Because art is the deepest of human expressions.

Revealing Genes

Jaume Bertranpetit

1976, Serengeti National Park, Tanzania. After hours of digging under the blazing sun, paleoanthropologist Mary Leakey and her team were relaxing and having a good time. They entertained themselves by tossing elephant dung at each other. While ducking a projectile, anthropologist Andrew Hill fell to the ground, only to discover a trail of fossilized animal footprints. Two years later under these first footprints, researchers found another set, belonging to a 3.6-million-year-old hominid. They had been made by an Australopithecus, an ancestor of ours. Previously discovered fossils of these hominins had yielded information about what our ancestors were like, but these footprints offered clear proof that they had walked upright, slowly, over the ashes that a volcano had dumped onto the rain-soaked savannah. Our evolutionary history has been pieced together over many years from milestones like this. Jaume Bertranpetit, however, doesn't need to toss around elephant dung to unearth footprints that reveal our distant past. He simply analyses the DNA retrieved from ancient fossils. Genetics can tell us not only what our ancestors were like, but also what kind of relationships there were between the various groups of ancient humans. Thanks to DNA analysis, we know that about 50,000 years ago, for example, we interbred with Neanderthals, another human species. As a result, anyone today who isn't from a sub-Saharan Africa population has 2% Neanderthal DNA. We also know that we were able to cohabitate with at least six other human species, which, today, we find unthinkable. What would it have been like to run into someone different enough that you wouldn't recognize them as one of your own kind, but similar enough to be considered human? Jaume Bertranpetit doesn't have an answer to that question, but his research lets us to ask it. And it is through questions like that one that our ideas about the world and humanity advance.

What do you think of Homage to the Elements?

Aesthetically, I think it's wonderful. What struck me most when I saw the mural was that no one had thought of putting the emission spectra on a periodic table before. Looking at it, it seems so obvious that they should be there! Plus, the work conveys the idea that there's an order in nature and that the periodic table isn't just a static representation but also includes the most dynamic part of the wavelengths we use to study the sky. Beyond that, those of us who study life often say there are four basic elements (carbon, hydrogen, nitrogen, and oxygen), and then we add a few more that are also necessary. In the end, though, you realize that there are so many involved. They all end up relating to one another.

> When it comes to studying life, one crucial idea is that species aren't fixed entities; they're constantly changing. That idea was developed simultaneously by Charles Darwin and Alfred Wallace, and some even say that it's the greatest idea ever conceived. In fact, it can be applied to many other fields outside the study of life.

Four years ago, when I left my role as director of the Catalan Institution for Research and Advanced Studies [ICREA], I took a sabbatical and went to Cambridge. At one of those dinners at a college where everyone puts on a toga, I was sitting next to a researcher who was working on classifying stars. And I told him that biologists stopped classifying things years ago. Maybe you can look at species like little boxes, but it's far more interesting to see how they're related to one another, to find out if they have common ancestors, and to understand why some have changed in one way and others in another. It's called phylogeny. And I told him that if he wanted to understand the stars, he should do the same thing. Biologists have had a tough time with fossils, but thanks to modern genetics studies, we've made a lot of progress because we've been able to mathematize the process of change. And that, I told him, should also be possible with the stars, because thanks to our knowledge of the emission spectra, like the ones you see in *Homage to the Elements*, you can determine the chemical composition of any star based on its light.

What did he say?
He suggested we should write a scientific article together.

And did you?
We did! It was really fun. The main idea was that in order to study two different things, whether they're species, words, or stars, you have to define what differentiates them. What makes two species different? DNA. And two stars? Their chemical composition. If for species, evolution is understood as the incorporation of mutations or changes in DNA, then for stars it is characterized by the generation of all the elements on the periodic table starting from the primordial elements from the Big Bang. It's an application to astronomy of the methods we use in evolutionary biology. It was the first instance of a stellar phylogeny as opposed to a classification.

When you explain it like that, it seems so obvious
that I'm surprised no one had done it before.
The funny thing is that it's directly related to the periodic table and the visualization created by Eugènia Balcells, which shows how dynamics are a feature you can find everywhere, not only in life but in matter.

Everything is dynamic, indeed. And all around
us, we can see that changes are happening
more and more rapidly. But are we evolving, as
biological beings?
There's an ongoing debate in biology that will never be resolved: if we could go back 600 million years and evolve again, would the same species emerge? The truth is, we don't know. They would probably be different from the species we know today, but there would no doubt be many clear indicators that would identify them as life on Earth.

Does that mean that evolution is always present and
that we humans are constantly evolving?
If we're trying to predict the future, we have to look at the mechanisms of evolution and how we alter them. One mechanism is mutation, a spontaneous change in DNA, which we already know has very few effects. The genetic drift mechanism, which means that some

characteristics end up being predominant by chance, can have effects on small populations, but in humanity as it exists today they're negligible. The same is true for the founder effect, in which a new population is established from a small number of individuals who carry only some of the characteristics of the original population. Because there are so many people, these mechanisms lose their effect. Then, there's migration, but because it's so widespread and so mixed, the effect is also diluted. So, the only mechanism left is natural selection.

Are we subject to it?
The truth is that the chance of natural selection acting is very low. It only works to increase the chances of survival or for better fertility. But we humans try to prevent as many people from dying as possible, extending life far beyond the reproductive stage. In other words, we try to improve survival for everyone who is born, and therefore, from that standpoint we prevent natural selection from acting. In terms of fertility, there has been a huge decrease in the number of children people have, so in that sense, natural selection also has very little opportunity to act.

So, we aren't evolving and we're biologically stagnant.
A few studies have detected a slight natural selection for greater height, but that took place thousands of years ago, not recently. Another thing to keep in mind is negative selection, which is also declining. Many of the diseases that used to kill us are now survivable. It's true that there are some diseases that cause death during the first few weeks of an embryo's development. But we know that if they get past the first three weeks of gestation, the probability of living to 80 is quite high. That might lead us to ask whether our trying to ensure everyone survives and can reproduce could have a negative effect on the genome of humanity in the future. The answer is that if there is a negative effect it will be very small, because now we've changed the environment to result in more well-adapted genes. I'm colorblind, for example. Is it more difficult for me to reproduce? Not at all. If I were a hunter-gatherer, it would be different: I'd starve to death!

I wonder if this lack of evolution can lead to problems. Because the world has changed a lot, but

our bodies and our brains developed in response to
the evolutionary pressures of living in the savannah,
which is pretty different from Wall Street.

The fact is that we aren't changing, and we won't. Culture creates
new tools, new adaptations, and new environments that individuals
can harness to live better. I think a more interesting question than
whether we'll change is whether we'll eventually get to the socalled
singularity, when machine intelligence is equal to human intelligence.
If that happens, how will we respond? Our intelligence has evolved to
survive, not to understand the world or alter it. Is that old-fashioned
intelligence necessary for survival today? The answer is no. If you don't
have it, others will help you survive.

Does that mean we are limited when it comes to
understanding or dealing with the future?

Of course we're limited! Compared to other animals, we're also limited
when it comes to running, flying, swimming underwater, etc. And
we're limited in terms of certain complex reasoning. It's clear that
machines can be invented that will understand many of the things
that we don't. That will come one day, but I'm not worried about it.
In my field of work, for example, we've applied neural networks and
deep learning to see something we could never see on our own: those
models have provided us with a series of potential evolutionary trees
to understand the evolution of our species. With just our knowledge,
we don't know which one is the right one; but the machines do. But
we can't know what process they've followed to reach that conclusion.
In that sense, it's clear that technology will bring about significant
changes and that they'll have major implications for our understanding
of the world, especially if at some point machines prove that their
intelligence is far superior to human intelligence.

Balcells says that light is the voice of matter.
Transferring the metaphor to your field, could we say
that genes are the voice of life?

It would be a little more than that because if we were to define life, we
could say that it's the part of the universe in which the instructions to
create it are included inside it. It would be like if the instructions to

put together a piece of IKEA furniture, in addition to being printed on a piece of paper, were self-executing and the furniture assembled itself on its own. Now, when we study the genome, although it tells us a lot about living things, there's still a lot we don't know. For example, we should be able to take a genetic sequence and, using a computer, bring it to life. That is, to create life *in silico* – virtual, but real. Today, we can compare sequences to each other, and we can tell whether they're from a newt or a lizard, or if they correspond to an African or Asian person, but we don't know how to make a genome work to create life. Not even the genome of a bacterium. But the information is there. Looking at the genome we should be able to see everything. That's why biology has its roots in DNA because it contains the foundation for how life works. So, going back to your question, it isn't that I can decide whether to study you or your genome. No! If I'm studying your genome, I'm studying you. They aren't two separate things.

> Everything is, in fact, in our genes. And because they can be analyzed, we've now discovered that we coexisted and interbred with other human species, which represents a paradigm shift.

Of course, because by the time modern humans crossed paths with a Neanderthal, they'd been separate for some 600,000 or 700,000 years, which is a lot. And the Denisovans, who we also interbred with, who were they? We say that we coexisted, at the least, with Neanderthals and with two welldifferentiated groups of Denisovans. Now, we have no idea if it was really two or three or 25. We don't know what the distribution of those hominins might have been, especially in Asia, before modern humans left Africa. But to study all this, we are now seeing that it's easier to reconstruct genomes than to search for bones.

> Going back to Balcells' work, you said it had great aesthetic value. Is this idea of aesthetic appreciation a human characteristic, or does it also exist among other species?

I have no idea, and I don't think there are many people who could know that. Most people who work in the field don't believe that other species are capable of aesthetic appreciation, but we don't actually have many ways to analyze that. Some researchers have tried to get chimpanzees to paint, which has been a real disaster. Jordi Sabater Pi used to say that what the chimpanzees painted looked like works of art, but I don't agree. What is clear is that the concept of beauty doesn't exist in nature but in our brains. Beauty is created in the brain. Sometimes, seeing a certain thing generates an internal reward in the form of pleasure. And we know that in humans, for example, making objects that go beyond mere utility is a very old and traceable activity. It emerged at some point a little over 100,000 years ago.

> Neanderthals also had the ability to appreciate
> aesthetics.

Exactly. And it is precisely that – since we know that this phenomenon did not exist until a little over 100,000 years ago – which suggests that attributing beauty to an object is an exclusively human faculty. Now, what is the adaptive significance of that behavior? I don't think we can know. But we can speculate. The evolutionary forces that have acted on the brain have either a social component or an ecological component. And I think the social dimension is the most important. Think, for example, of our ability to recognize faces, which is extraordinary. What ecological need was there to do that so well? Well, it must be very important, because it's one of the foundations of social relationships. In primitive societies, recognizing faces must have been crucial to knowing who you had treated badly, who owed you something, or whom you owed something to, who needed help, etc. In other words, it was essential to situating each individual perfectly within a complex map of social relationships. That must have been fundamental.

> That means that, in a way, art makes us human.

You could say that, yes. Although our artistic nature is a characteristic that we don't really know how to explain from the standpoint of biology. It may not have an explanation, because it isn't always possible to explain every detail of an organism based on a specific function. It could be that some type of intelligence that was very important in our evolutionary history resulted in artistic creation and aesthetic appreciation as a byproduct.

Homage to the Elements has a simplicity that serves as a very apt representation of the characteristic of art and science of saying as much as possible using minimum resources. This is true in science, in the case of theories that are increasingly more general. And there is a similar feeling in the art world when you come across a piece that contains a whole universe. Do you see art and science as two separate worlds or do they have things in common?

Balcells' piece is a very clear example of working with exclusively scientific tools and turning them inside out like a sock to reveal this enormous beauty that was right there all along. I think a small part of the art world is trying to look at science, and there is a very small part of the world of science trying to look at art. They are small portions, but they should be nourished.

Did you say small and very small for a specific reason?

Because there are sectors of the art world that use science or, rather, technology, even though they aren't aware of it. Without even getting into photography or film, painters use pigments, an area that is pure chemistry, and there are artists who use a lot of technology. Technology is the link between art, science, and everyday life, but the dialogue with the world of science is often limited.

Aren't there many scientists interested in art?

Not so many. Some scientists paint or sculpt, but that isn't particularly important because they probably aren't great artists. The important thing is for there to be conversations and intermingling between the two groups. It would be great if there were an artist at each research center who could help us think outside the scientific mechanisms we use every day. Because scientists are good at making deductions, but we also need the induction that art can offer. I'll always remember the time I spent playing with the actress Anna Lizaran. We played a game based on colors and I would think and think, and she would beat me in just seconds.

Pure intuition.

Exactly. And what is the path to intuition? We don't know a lot about intuition. Sometimes it will hit us while we're taking a walk or driving home, but it isn't a mechanism we're trained to use. Artists, on the other hand, do. That's why I think there should be more works of art like *Homage to the Elements*. There need to be more things like it.

Dancing Life

Cesc Gelabert

I'm not a great dancer, I'll admit. I don't think it's a question of coordination because I would say I'm fairly competent at more than one sport. And it isn't because of a tin ear either, because I can play a few instruments relatively well. I have the feeling that it has more to do with a blockage, which a lot of people probably experience, that cuts off the connection between my brain, with all its emotions and rational thoughts, and the rest of my body. But that inability, which at this point could be called chronic, and the frustration it causes have given me a certain propensity for appreciating the harmony that can exist in any movement. Ultimately, someone who is a good dancer isn't that different from an acclaimed athlete: Leo Messi's dribbling, a pole vault by Yelena Isinbayeva, Roger Federer's backhand or Steffi Graf's forehand, Eliud Kipchoge running the marathon, or Simone Biles doing a triple-twisting double backflip. They are all invitations to contemplate a certain kind of harmony that not only contains the intellectual milestone of achieving a goal, but also a beauty that reconnects us with our bodies and, at the same time, amazes and comforts us. However, it could be argued that the beauty in sports is something spontaneous. It isn't a prerequisite; it simply appears in the process of pursuing a physical goal. In dance, on the other hand, although it is a very emotional discipline, beauty is constructed with complete awareness. As a result, the levels of admiration it can elicit can, potentially, be very high. I have seen this, for example – often unexpectedly – in quintessential classics like the Mariinsky Ballet of St. Petersburg or in performances by the contemporary dancer Hiroaki Umeda. And, of course, in Cesc Gelabert. Because through four decades on the stage, he has built a discourse that is intellectual and physical at the same time. When he talks, when he walks, when he dances (obviously!), but also when he's just sitting in a chair, he radiates a kind of harmony that can only come from a radical mind-body balance. Everything his body does is founded on an intellectual discourse and, at the same time, his entire logical argument begins with a movement, a gesture. And this consideration provides the observer with that aesthetic pleasure, which the writer Vladimir Nabokov describes as unparalleled because it doesn't depend either on the brain or on the heart but on something that connects them. I wish I knew how to dance.

There are pieces by Balcells, like *Frequencies* or *Universe*, that have to be perceived by something more than our senses. They offer a very physical experience that almost needs to be felt with our bodies, which is a dancer's medium of expression.
We're condemned to dance forever.

Why is that?
Because we can't avoid experiencing life through an inhabited body. I always say that dancing means inhabiting your body with your emotions and your mind. It isn't just a physical action. For me, it's an action that sums up a certain spiritual search, which I have always engaged in from an artistic standpoint and which leads into the idea of a waking dream. When an audience receives a dance like a dream shared in a waking state, it closes the loop. And, conversely, if the person who is watching doesn't perceive it as art, the art doesn't exist. But dance goes beyond that. Life is just fragments of choreography with meaning. One of the most astounding choreographies is the one children create when they're learning to walk. It's incredible! When someone hugs you, you have to be able to understand whether that embrace is genuine. What someone says about it isn't enough. And that is essentially dance. It doesn't make sense to be cut off from your body.

In that regard, dance is useful in a lot of ways.
Absolutely. And it's such a shame that we don't use it for communication, health, expression, wisdom... Lately I've been holding a lot of classes for teachers and social educators. I also try to get involved in the world of work. I would love for dance to be at the core of work and education. I would love for modern corporations to use it to improve relationships with their employees, and for schools to use dance as a way of teaching math or physics. Plus, dance has another advantage: it isn't ideological. Very different kinds of thought can be inscribed in the body.

And yet, it's a minority discipline.
Unfortunately, there is such cultural ignorance when it comes to movement in our society that our seniors get to a point where they

can't even bathe themselves. And that means that the corresponding art – dance – is in trouble. I imagine that prehistoric tribes like the ones that painted authentic wonders in the Chauvet Cave must have had a very different relationship with their bodies and movement. I advocate that every form of culture, from sports to robotics, should include emotions, body and mind.

> And does dance get the recognition it deserves in the
> cultural world?

A diplomat friend of mine used to say that dance is one of the few arts that a cultured person can admit they know nothing about. Most people have never seen the equivalents of a Picasso in the dance world: for example, Merce Cunningham or José Limón.

> Like some pieces by Balcells – *Universe* **or** *Frequencies* –
> dance has a performative nature that makes it
> ephemeral.

I've always said that there are two kinds of dance: the kind that is and the kind that isn't. Just like there are two kinds of writing, two kinds of thought, etc. Historically, of course, there is a wide variety. There is a cultural heritage that evolves over time. If you look at classical dance, now it seems very organized, but if you look back at Anna Pavlova, the first person to improvise a flat toe box for her pointe shoes, that was wild. Like someone doing hip-hip or performance today. The history of dance is complicated because you can't see it. To draw a parallel with the other arts, it would be like if all the monuments or paintings disappeared and we were left with just their descriptions. The description of a Rafael, for example.

> *Homage to the Elements* is different in that regard:
> a mural that remains. What are your thoughts on
> the piece?

Not long ago I was reading a book by Oliver Sacks, who had a longstanding fascination with the periodic table. What could be more intriguing than something with the ability to describe all the elements in nature? In that sense, *Homage to the Elements* is a magnificent example of an internal and external vision of the harmony of the cosmos. As a viewer, it's beautiful because you get the feeling that you're connected to the world. It's like the Theatre of Memory in the Renaissance. It seems like the world is within your reach. And at the same time, you have the feeling that you don't fully comprehend it, that you're dreaming, that your imagination is taking flight in the face of what you're seeing. It has visual elements that don't demand anything from you. You just have to receive it like a child. Looking at all those strips of color in itself is already incredible. Plus, you can understand it, and relate it to yourself. And if you danced it, if you danced all the elements, it would be like dancing all the possibilities that dance gives you, from the fast elements to the slow ones, from the highest energy elements to the lowest. I've gone to great lengths studying energy in the body. And this would be one more way of doing it.

> Frank Zappa said that talking about music was like dancing about architecture. Could you dance about one of Balcells' pieces?

I'm a dancer because I can relate dance to anything. I can dance whatever you want, from a play in a football game to a meal. And I could also dance those spectra, which emit such an amazing visual conjunction.

> You were saying that *Homage to the Elements* is a vision of the harmony of the cosmos. Do you think Balcells' work has the ability to look at everyday things and see beyond them?

Eugenia is like a natural explosion. She hardly breathes. And I think that's fascinating, because she has a wonderful relationship with her body: a body that doesn't breathe, but which is entirely present, hyperventilated as it is. And that's why the work she produces is so subtle, but in the background she's so excited and she has a body. The intellect by itself is nothing, just like the body by itself. I think Eugènia is a subtle and very interesting artist, because she is a part of a broader milieu that focuses on understanding the meaning of life. And she does it very faithfully, and with a lot of care. She ended up entering a world that is half conceptual but very experience-based, and she has found her own, very unique territories there, like that of the spectra. I understand why she's fascinated by them, because it's a question that shares all the characteristics that make up her artistic vision, which revolves around a way of looking. An excited, animated and deliberate gaze. And a gaze that is anchored in a place. She isn't a painter who paints physically, she paints with her eyes.

> And in that painting with her eyes, light plays a
> fundamental role. How importance is light in dance?

The most important thing in dance is presence. Sometimes I'll talk to someone and afterwards I won't even remember the color of their eyes, but I'll remember their presence. Even if I've only seen them on a phone screen. That presence, to an extent, is the body's light. And that connects with someone's interior world, with that idea of dance as being inhabited, spiritually. For example, when you imagine a figure of the buddha, you imagine him with a body made of light, which, in our immediate universe we can understand as something that isn't purely physical. On the other hand, when I'm working on a choreography, the first thing I see is the light that surrounds it. I've always like working on lighting design with the designers. Ultimately, my movements are different depending on whether the light hits them in one way or another. If the lighting isn't right, the performance falls apart, because the lighting has a direct connection with the emotional aspect. I think landscapes affect us, above all, because of a luminous quality, and that's what we connect with emotionally. There are a lot of interesting subtleties in Buddhism in relation to light, but light as a metaphor can be found in all cultures.

> Buddhism has been very important in your life,
> hasn't it?

Since I was 14, I've always taken spiritual exercises very seriously. I studied Christianity and then many other religions and cultures, but the teachers I liked best were always dead. Then at some point, I came across a book by Geshe Kelsang Gyatso and when I read it, it seemed like everything else I had read but from a Buddhist perspective, and it was coming to me from a man who had a mailing address. I became a follower and I've continued meditating ever since. And it has been a lifeline, because it gives me a perspective that includes the entire past, the entire future, everyone, the entire present, and everything that is most profound. It's a cosmic perspective that has helped me find a place for myself. There was a point in my life, years ago, when I went through a serious crisis. Art was really important to me, but I was seeing things in it that I didn't understand, that disenchanted me. I started to feel unhappy and lost, because I couldn't share something that was important to me with people that I respected deeply. Through Buddhism, I understood that everything depends on the mind, on our perception, and that helped me find my place. It gave me perspective and I could relativize things a little more. Today, anything can be art, and our generation has worked hard to make that possible, but now it's become harder to distinguish between what is art and what isn't. It's very subjective.

In the case of dance, do you know what art is?
When I talk about inhabiting our bodies with our emotions and our minds, it's something ongoing. Everyone does it all day long. But that doesn't mean it necessarily has to be art. If you stop and think about it, a lot of our cultural heritage – from Tai chi to football – is a manifestation of that inhabiting the body. And all of that has the possibility of becoming art. But, in the end, what's important is culture; we aren't what's important. When I do something that works, it's because the art is *in* me, as the ancients would say. When the art is *in* me, I can do something worthwhile, and when art and life truly come from within, then they can be original and valuable. Eugènia remembers me dancing from decades ago, and I remember her artwork. That's what interests me, that contact, reaching past the present moment and our immediate surroundings.

The Importance of Refusing to Choose

Sunetra Gupta

Stories about double lives have always captivated me. The diligent post office clerk who, when the sun goes down, strains her vocal cords screaming over distorted guitars at a dive bar downtown. Or the reliable bank manager who is also the star of a drag show at a rundown cabaret. Sunetra Gupta doesn't lead a double life in that sense, but if we extend the metaphor a little, we find a fundamental similarity. Whereas in the case of the postal worker we might ask why working at the post office should interfere with her passion for rock and roll, in the case of Sunetra Gupta, the question is, why should working as a scientist at one of the world's most prestigious universities prevent her from writing internationally acclaimed novels? You might say that it could be very hard to do both things well. And you would be right. But she does it. And I would say that's because she has never given up on either of those things. She never paid any mind to an educational system that demands we choose between science and the humanities at an age when no one quite knows what either of those things are. She has always ignored both the complacent "I'm an English major" and the equally fateful "Sorry, I studied science." Her example shows us that a double life – or to put it in more tactful terms, a more expansive life – is possible. And those of us who have always felt that being forced to choose is a limitation, we see in figures like hers all the markings of a butterfly that has just emerged from its cocoon. A possibility of flying that should not be ignored.

> What did you think of Eugènia Balcells' work the
> first time you saw it?

I first saw *Frequencies* with my daughters, who were nine and twelve
at the time, and all three of us, each in our own way, felt completely
engrossed by the installation. The girls lay down on the floor and were
amazed by the atmospheres Balcells created with light. As for me, I
remember being very moved, and I felt that it echoed the beauty of
work by certain abstract expressionists.

> As a scientist, how did *Frequencies* and *Homage to the
> Elements* speak to you?

I think they contain one of the richest truths of science, which is that
the simplest elements can be combined to bring about a very wide
range of phenomena, from physical phenomena to our emotional
states. And, in a way, that's where science and art come together. The
traditional view of science is that it is a reductionist activity based
on simplifying things using a very specific language. According to
that view, the opposite is true for art. But in reality there is a point of
connection between the two, exemplified in chaos theory, which says
that a wealth of patterns can be formed drawing on very simple rules.
And those patterns aren't just noise. An important distinction needs
to be made between the richness and complexity of the patterns, on
the one hand, and noise, on the other. Most of us want to live our lives
to the fullest, not just according to simple ideas and rules. We want to
push past our limits. That's what humans do; we push our boundaries.
But we run the risk of entering a world full of noise, and there's no
utility in that. I think a lot of art is made on that border between order
and noise, which is where chaos sits.

> Centuries ago, science and art were more closely
> related. Today, people see science as a useful activity
> that has little to do with the idea of beauty.

You're right, and that's tragic. In fact, what attracted me to both
science and art is the deep connection between the two activities,
which are like two aspects of the same spirit of investigating life. And
both of them have elements that are enormously beautiful, sometimes
in different ways. It's truly sad that their connection has been lost.

What is particularly sad is that market pressure has led to science being valued only in terms of its usefulness – a usefulness that is also very much determined by questions like "Will this increase our GDP?" Plus, now the same considerations are also being applied to art, and that's something that really scares me.

> You're also familiar with other pieces by Balcells, like *Color Wheel* and *Labyrinth*.

Those pieces again combine a kind of sensitivity to objects, colors and memories. Throughout Balcells' work there is a sense of how we relate to those elements, things that brush past your face or things that you watch as they change before your eyes, and that process of change is captured in the most poetic way possible in her work. It's wonderful. There is also a wonderful element of chance, a celebration of chance. You find something in one place, you take it out of context, like cutting a circle out of a magazine, for example, and all of a sudden it becomes something completely different. The possibility of creating a meaning for life, a full life, is in all her work, through the possibilities inherent in objects that we find in our daily lives.

> There is also an idea of harmony, a harmony that we may have lost touch with.

A very clear example of this broken harmony is how the world has dealt with the pandemic. From the beginning it wasn't considered a global problem but rather a local one. Many countries closed their borders to keep the disease out. There was a generalized feeling of "So long as we're OK, who cares about everyone else?" What I found most shocking about the pandemic was the levity with which everyone accepted the consequences of closing ourselves off and isolating ourselves, which runs counter to human nature. I've been struck to see many scientists thinking in that way, considering only their own country, without taking a global view and without being aware of what it means to be alive and to be part of an international community. I think, in reality, that there is a very strong connection between all of us. I think that if all humanity were instilled with the celebration of harmony and the fact of being alive, we'd be better prepared to handle situations like this.

What can we do to transmit those values?

We should invest much more in making art accessible to the general public and, in particular, to all schoolchildren. We need to expose students to things that are a little mysterious for them, that expand their minds and help them celebrate life and get in touch with its meaning. Things that give them a vision of life where the risk of a pandemic is something that we have to contain and live with from an international standpoint. An artistic education can prepare you for events like this, where you're forced to make very difficult decisions about life and death.

Is our education too focused on learning useful things?

There is a strong emphasis on training professionals in the narrowest sense of the term, a type of education and compliance that doesn't leave any time for developing other abilities, like artistic abilities. If people want to succeed in a field, they have to devote all their time to it. And governments seem inclined to promote very specific activities that are aimed at incrementing some type of wealth.

It is also true that there is much more scientific knowledge today than ever before, and becoming an expert in anything takes a long time.

Of course. I'm quite familiar with the limitations of the system. Becoming an expert in a field takes a lot of time, and that's just one key issue that we need to keep in mind. At some point, people need to be able decide whether they want to spend five years working hard to become a surgeon. They may dedicate certain periods of their lives to a specific kind of training if it's necessary, but putting one thing above all else neglects the richness of our minds. In the end, it's true that we do need people who want to spend all their time writing computer code, to give just one example. We have to accept that not everyone will need to embrace the world of art, but I think we should ensure that schoolchildren have the chance to see for themselves all the different ways they might live their lives as adults. Do we really have to choose? Some of them may want to be excellent tennis players or groundbreaking neurosurgeons, and that's great.

Those are good choices too. But in the end, it's more beneficial for the population when everyone can pursue different interests and combine their sensitivities toward art and toward science. And our goal isn't necessarily to understand everything, but to let ourselves delve into the mystery.

What do you think people who aren't knowledgeable about science can see in Balcells' work?

Precisely, they can see the beauty of science and move away from understanding it as a purely utilitarian activity. There is a lot of poetry in the fundamental elements of science, which is also accessible to anyone, even without a scientific education. And when that poetry is filtered through the sensibilities of someone like Balcells, it creates something very special, which scientists also appreciate and recognize: in this case, the building blocks of life and the universe. But what's more, when there is that filtering through the sensitivity of an artist, in a way, it reveals the potential of those ideas beyond their scientific utility.

The Art
of Teaching

Stephen M. Noonan

The future has always been a mystery. But today it may be even more unfathomable than it was just a few years ago. People predicted there would be flying cars, but not electric scooters. And no one imagined that we would be walking around with our eyes glued to tiny screens. How, then, can we prepare ourselves collectively for the unknown? Education is definitely one possibility. Educational policy is a country's most significant commitment to the future, but how can it be approached to deal with the uncertainty caused by the extreme acceleration of technological development? At the New York high school run by Stephen Noonan, the answer is clear: by incorporating the arts into everything they do. It's curious to think about preparing for a future that is more uncertain than ever before by looking to recover certainties from the past, when categories were less clear-cut and, for example, music was understood as an area of mathematics or geometry was included as part of painting. Stephen Noonan talks about all this with a contagious enthusiasm that conveys, above all, a radiant optimism. As they say, educators have to be optimists. You can't be a teacher without the hope that what you're doing today will have a positive impact on the future. Beyond optimism, the uncertainty we're facing can only be tackled with resilience and creativity. And Stephen Noonan has an abundance of both.

How do you approach education at the Maxine
Greene High School in New York?

Our high school was founded to continue the work of Dr. Maxine Greene, which was founded on the idea of integrating works of art into the educational curriculum across the board, with the goal of helping students make deeper connections. The way we do that is with artists who work side by side with teachers, offering very valuable knowledge about music, theater, and the visual arts. In that sense, over the years we've seen that it's very easy to make use of a work of art in the humanities, in a language or history class, but that it has always been a challenge for teachers to find works of art that connect meaningfully with the sciences and math. They often work in the opposite direction: they focus first on major ideas from math and science and then look for a work of art that aligns with their curricular requirements.

And did that change when you found out about
Homage to the Elements?

When we discovered that piece, everything made sense. Because it creatively represents a tool that students use in science classes, the periodic table, but in a way that allows them to make that artistic connection. When they look at it, they see a representation of the periodic table that they can connect with in other ways. Our high school has six buildings, and we have *Homage to the Elements* installed in the lobby of one of them, right next to the entrance to the library. Our hope is that students from other buildings will also come and see the work as if they were going to an exhibition. Right after we installed it, we invited teachers to come see it and start thinking of ways to integrate it into their programs. Having this piece here is a wonderful gift to our entire educational community.

What were your first impressions of the piece?

I have a background in the humanities; I used to be a history teacher. Mathematics and science are not my subjects, but as a high school principal, I work with all the teachers to ensure that we provide a quality education in all areas. The first time I saw *Homage to the Elements* I was very impressed because it gave me a very different experience of the periodic table than I had had until then.

My previous experiences had always been very flat: tables with numbers and letters on a sheet of paper, in a notebook or hanging on the wall, often in black and white, which teachers used to explain the elements and students used to learn chemistry. But when I saw the periodic table represented by light and color, a world of possibilities opened up for me.

Like what?

Now, for example, we work with digital art a lot: students make three-dimensional pieces on the computer, they paint them using the colors from a palette, and they can move them around in space. And I thought that using *Homage to the Elements*, students could look at the elements in their most basic form, which is light and color, and connect them with their creative experience. For a student like me, who was not very good at science and math, that might have helped me to connect with the periodic table in a different way. Because if I had seen the elements represented by light and color, I would have been able to access that tool, the periodic table, in a new way. And that opens up a lot of profound possibilities for teaching and learning.

What impact has the piece had on the school?

The pandemic has stalled things a bit, but the challenge is always to ensure that the works of art are being incorporated into the teachers' lesson plans in a meaningful and appropriate way, in keeping with the learning requirements that students must meet. They aren't something we just hand on the wall and that's it. We don't want the students' contact with the artwork to be an isolated experience, but for it to be consistently integrated into the curriculum and to be meaningful. In that sense, we're trying to get students to connect to the piece before they start working with it academically, so that they've already been exposed to it, are familiar with it, and know what they're looking at. Then, when it comes time to use the periodic table, the piece is another resource at their disposal. And the teachers from the various subjects have all worked toward that. Since the piece hangs in the lobby and it's huge, students pass by it every day as they go to class or to the cafeteria. As Maxine Greene put it, the power of art is that students are exposed to it and find it familiar, while still unexplored, so that when they start working on the piece more deliberately, it becomes a very powerful resource.

Why do you think art can be so useful when you incorporate it into the teaching curriculum?
Mostly because of an idea that Maxine Greene also explained: the power of a work of art doesn't lie within in, but in the relationship that an individual establishes with it. If students have the opportunity to access a piece of art and are allowed to connect with it, they will develop a relationship with that work, and that will give a deeper meaning to their educational experience. They will be able to talk about it and have a richer experience in class and throughout their academic career. Because, really, it doesn't come down to appreciating art, it isn't about art for art's sake, it isn't even about art being beautiful or important. It's about giving students the space to experience works of art and discover what is meaningful to them. When we do that, the students are empowered, their curiosity is piqued, and they start asking themselves questions. And teachers need to be able to encourage, support, and guide them in this dialogue with the work. We are also lucky that there are always multiple entry points to a work of art. You can look at it from the point of view of the composition, the story it tells, the color, and so on. The role of the teacher is to recognize all these approaches and act as a guide in the process. And over the years, we've seen that these experiences are very enriching and meaningful for students. They remember them. And they remember how the work made them feel and whether they connected with it. It's also interesting to note that we work in New York City with high school students who have certain difficulties, like problems with reading or basic math. Our goal is for them to be committed to academic work, in which they haven't been successful in the past. In that sense, I think that art can awaken something in them that lets them connect with the piece, the teacher, and the subject. That's why I believe that a visionary piece like *Homage to the Elements* provides the perfect opportunity for students to have that experience.

Your high school specializes in integrating art into the curriculum, but do you think works like *Homage to the Elements* can be useful at any high school around the world?

I think so, especially in the sciences. Because in science we ask students to develop hypotheses, to think creatively, to be motivated, and to solve problems. Since *Homage to the Elements* presents the elements in ways that students haven't seen or thought about before, it raises questions that generate a deeper motivation, regardless of the subject matter they're working on. And that connection is what we teachers are looking for. We want students to push the boundaries of content. Maxine Greene always said that answers aren't as important as questions. Because one question leads you to another, and through that process of questioning, students come to live deeper and more meaningful experiences.

How have students reacted to the piece?

So far, they have done a couple of preliminary presentations on how they've worked with *Homage to the Elements*. And what we've seen is that they've made very concrete and tangible connections between the work and what the elements are in themselves. We had a couple of classes where each student took a single element and immersed themselves in everything that characterizes it and how it relates to the other elements. Then the same students created representations of how they see the element in the world and presented them in front of the mural. It was amazing to see how deep they had gone. And I think they were able to do that thanks to the fact that *Homage to the Elements* is so visible and, in a way, very real and yet very different from the classic periodic table.

Over time, educational content has become fragmented and disconnected. Physics and history, for example, are taught separately although they are intimately related in the real world. Do you think that the kind of integration you do is where education should be headed in the future?

For me, it is very clear. Over the years, we've seen that the experiences students have with art build on each other, and each new experience enriches the previous ones. And we think that's very important because we work with your everyday students. We don't run a program for highly gifted kids. We're simply committed to giving our students the opportunity to work with pieces of art. And I think those experiences resonate in different ways than traditional classes. As opposed to having the students learn something, take an exam, and then move on to the next topic, we try to do something a little more complicated: we make sure all the teachers are familiar with the works of art and their significance for each grade level, and then when the students make connections, the teachers support them in their work so they can see that the pieces are related to all their subjects. That leads them to ask more questions and make even more connections with the art. And the more experiences they have throughout the curriculum, the more powerful the educational experience will be.

> What impact do you think *Homage to the Elements* can have on an adult who has studied all the different subjects separately at school?

I hope it will enlighten them and help them overcome the barriers that have been artificially imposed on them. In the past, education was very linear and prescriptive, very directed. But now, in the 21st century, young people don't live in a linear world anymore. Everything comes at them in different times and places, and they need structure so they can make the necessary connections and extract meaning from all those intense stimuli and experiences. And I think these preliminary experiences with works of art can do that for all of us. Maxine Greene's philosophy was very deliberate in helping us understand that not everything we live through is a linear experience. And one of art's great virtues is that you can always go back to it. And when you do, you see it with new eyes. Talking about the periodic table as a work of art as opposed to a repository of scientific information creates an experience that can live on in people. Someone who has worked with it in that way will see the periodic table differently when they look at it in the future. And that's why I think our work can be so powerful because we're creating spaces that allow those experiences to happen.

> And what if a student doesn't like the work or doesn't connect with it?

No problem. Not all works have to speak to everyone. That's where the work of the teacher comes in. I don't mind at all if a student tells me they don't like a work of art, as long as they can explain why. At the very least, teachers fight to keep the students open to the experience. That's why I think *Homage to the Elements* has so much potential because it expands the language we use to talk about science. A lot us think or have been told that we aren't good at something, but if we talk about science the way we talk about art in class, we can open doors and opportunities for students who think they aren't good at science.

Is there a risk of distancing students from teachers in these fast-paced times?

That has been the major issue since we emerged from the caves. Young people are always one step ahead of their elders. But teachers have to keep up to date. Our job is to help students make sense of their own experiences, whatever those experiences may be. Good teachers ask good questions, they make you think. Teachers shouldn't be standing at the front of the class telling students what they need to know and what's important. They should accompany them on a journey of mutual discovery. And I think that happens more often when the experience of art is introduced because in art there are no fixed answers. Two and two is four, that's obvious. I don't care what you believe, or what you think, or what you want it to be. Two and two is four, period. But what does four mean to everyone? That's not so easy to answer. If teachers are able to ask good questions, the educational experience is very different from just taking an exam. In fact, we're preparing students to live in a world that doesn't exist yet. We're preparing them for a world that will be radically different in 10 years. And that's very intimidating, but at the same time, it's very liberating. It means moving towards the unknown! In the end, we want students to be motivated and connected to the world. If we manage that, they're likely to maintain that throughout their lives and succeed in that new world. That's why it's so important to question things and imbue them with meaning. For Maxine Greene, asking questions that lead to more questions was more important than finding answers. Because there are questions that have answers, but there are also always questions that don't. In the long run, life always takes us places where there are no answers.

An Immaterial Material

Marta Llorente

It's such an integral part of our lives that we can't imagine a world of darkness. We need it because it has always been there, and as the products of time and chance, we've evolved to take advantage of it to capture the basic information about our surroundings that lets us interact with them. Animals that live in the depths of caves are blind. They don't need eyes for anything. But in our case, we not only "see the light" and relate enlightenment to knowledge and clarity, we also need light physically. If we don't get enough sunshine, we can become vitamin D deficient and feel weak and sore, and our bones can get brittle. The ethereal substance that is light structures us physically. But that's not all: when we enter a cozy house, just as the late afternoon light is streaming in through the window, having come through a thicker atmosphere than at noon ridding it of its bluish hues, it gives us a sense of warmth that makes us feel safe. This need for protection – which light satisfies so well, and which is strictly connected to our evolutionary history – would not be possible without architecture, without the manipulation of space using certain materials, intended to create shelter (protection, again) from the elements and to harness the light. Marta Llorente has thought and written extensively about light, not only about how we use it to feel safe and secure, but also about how we protect ourselves from it and, above all, how we use it as an intangible tool for creativity. A tool that was used by cave painters and that artists like Kazimir Malevich and Mark Rothko distilled to the delight of not only experts like Marta Llorente, but anyone with a pair of eyes in their head.

> What are your impressions of Eugenia
> Balcells' work?

I really like her work because her pieces are alluring. There's beauty in her way of doing things. I don't know if she's fully aware of it, but for me she's a redeeming presence, like a brilliance, an aura. Although she does very unusual things, she never loses sight of that balance, that harmony. I like that she offers a contemporary analysis while making a small concession to an element that accentuates the value of what she does. She doesn't just work with an idea; she accompanies it with an aesthetic.

> Do you see any connection in Balcells' work with a
> deeper reality that might be described as mystical
> or spiritual?

I don't practice any religion, but I understand spirituality as a longing, a nostalgia for a possibility that it isn't easy for me to find in life. It's a ceremonial founded on a delicate approach to things. Eugenia is very passionate and very mystical. I remember that, when I met her, my father was dying, and at such an upsetting time for me, she offered me a very good explanation for death. I think she has a kind of antenna for those things. And I do find her art to be spiritual, and scientific. But science has a close connection to spiritual questions. There are people who find religion after working in physics. And I'm not surprised, because everything is so strange...

> Could this nostalgia be caused, in part, by
> the gradual abandon of religious identity in
> modern societies?

I suppose so. I've worked a lot with Eugenio Trías, who wrote about religion, and I remember he was engaged in a desperate search for what we've forgotten with all our studying and knowledge. George Steiner wrote a book called *Nostalgia for the Absolute*. We have swept away religion, deservedly so, because sometimes the institution of religion and the things that are done in its name are aberrant. We needed a clean slate and we turned to materialism. But now, with the pandemic, with so many people dying alone, we've had to reconsider the carelessness with which we do so many things now.

Homage to the Elements and many of Balcells' other
pieces are based on light. As an architect, you've said
that light is one of the materials used in architecture,
just like iron or brick.

I say that light is an immaterial material. In that sense, it's a very
different material from the rest, because it's material but it isn't
matter. And that makes it very complicated and at the same time very
important to architecture. Plus, light is disastrous for the habitat.

Why is that?

Because we have to modulate it if we want to survive. We can't handle
the light outdoors. We couldn't survive in the Mediterranean, we'd be
scorched to death. We have to protect ourselves from light as it reaches
the world. In that sense, architecture is a poetics of light. Because the
materials that we can touch and that generate a tactile experience
create a memory that may even be muscular or sensitive, but light
escapes all that. What's most important is that light in architecture
is hard to tolerate. Light dazzles us; without architecture we couldn't
have withstood it. We protect ourselves from it, but at the same time,
we need it. And so, it's up to us to play with it. For me, architecture
is play. Light is a functional material that we play with because we
know how to create it technically. Not just since the invention of
incandescence, when street lighting was introduced and light was
democratized. Before that, there was always the question of created
light, of flame.

The 19th-century English scientist Michael Faraday
wrote a book called *The Chemical History of a Candle*
in which he analyzes this man-made light from a
point of view that is scientific but also very poetic.

The book is based on a series of lectures he gave, which blew everyone
away. He talks about the quality of this artificial light – because it
is artificial, in a way – and that means that architecture is just the
mysterious box in which we experience this creation of a light that we
make ourselves. Exposed to the elements, not even the protection of
fire would be sufficient. And in relation to all this, there's the absurd
myth of the cave. The cave functions as a sanctuary, as an exceptional

refuge, but the first architecture was a hut. And a space for fire was immediately made within that architecture, because it's an absolute necessity. Plus, architecture also serves as a sundial. I've learned to tell the time in my house without looking at the clock because there's a dialogue between light, space and time. And if we look at the periodic table, there's this mystery that the elements have in their spectra, which, ultimately, explain light in their own way. The periodic table itself is a wonderful thing. It's thrilling to know that there were empty sections before some of the elements were discovered. That means there's a coherence to the whole.

> In your book *Construir bajo el cielo. Un ensayo sobre la luz* [Building under the Sky: An Essay on Light] you talk about the significance of light from the point of view of biology, poetry, religion and the visual arts. In the art world, light has also been used almost like a material.

It's always worthwhile to do a second reading of any classical painting. Caravaggio, for example, never painted with natural light; instead, he did it in a very dark space so he could make his own light and create an impression of naturalness. It resulted in a very contradictory effect – let's not forget that he was painting during the Baroque – where, in reality, everything is false. The Impressionists, on the other hand, looked for light outdoors because they didn't want that ethereal artifice. The abstract painters, for their part, are incredible. One thing I'm really interested in is what happens when a painting is stripped of any narrative, because then the narrative is just light and color.

> And what happens when the narrative of a painting is just light and color?

One case I've studied in depth is that of Malevich. When he paints *Black Square*, he says that it won't be a painting but an icon, because he conceives it as a sign to erase the memory of the world and introduce the perception of a hypothetical world that won't even be real. And to lay the foundations for a new world, you have to go back to the fundamental elements of this world, which are the colors. Then he creates a series of pieces that he releases into a new world, and they are pieces, precisely, that use basic colors. And that isn't a coincidence; it isn't an aesthetic whim. Then in 1919, with the Russian Revolution and World War I already underway, he had developed the colors and he does the series *White on White*. He also went through a terrible experience: before his death, in the time of Stalin, his house was searched and his work was burned; it was later recovered but he never knew. When you see his pieces at the Stedelijk Museum in Amsterdam, and you look at the *White on White* series, you realize that what he's doing actually is very simple. What you see is a nuanced white, a small shadow of a white on white, the last thing that is visible of the world he created. It's a fantastic story. And when he gets to that point, he says: "Painting is over." But he doesn't end it with black, but with white. He has run the full range of colors and arrived at light.

What can be painted after that?
Well, he says: "Now we're going to move into a different dimension." And he starts designing Suprematist architecture using models. And he plays with light and creates a completely anti-functional architecture. Malevich was a visionary who took his inspiration from the abstract expressionism of Mondrian, Kandinsky, and everyone who was breaking with the object at that moment. When I look at the colored lines on the periodic table from *Homage to the Elements*, I can't help but think of abstract expressionism and Rothko, who also had a very tragic life. I have a vivid memory of the effects of his paintings around a single person, which I experienced at the Miró Foundation. And I remember the space very well because I always look at museums as spaces. That's why I think paintings should be in museums. They speak to you based on how and where they are hung. And I think Rothko is looking for a vibration that has something mystical about it, but he's probably intuiting things about the basic nature of matter and its expression. Rothko is a mysterious painter because he elicits an almost physical effect.

What is that effect like?
I feel an experience with color, it's a very strange thing. I can't explain it. I don't know what color is for you, it's something intangible even in language. When I say "red", what exactly am I saying? In reality, I'm

not really saying anything. I'm not measuring the red. I guess if you measure the wavelength of light then there is a saturating red, but our perception is so conditioned by the relationship between color and objects in life that we can't separate it. We can't dissociate red from blood, for example. Color is a concept. And, in that sense, I can't know what red is for you.

> We've talked about the importance of light and color in painting, but what role does color play in architecture?

Architecture is color too. And we often forget that, because, in the Western world, nearly everyone associates the architecture of the Modern Movement with white concrete. But that's not true. Le Corbusier painted the walls in different colors. He was heavily influenced by painting, and color played a very intimate role in his interiors. He wasn't aiming for a spectacular expression of color in architecture, but he was looking for that vibration of the relationship, and almost no one looks at how he used color in his interiors. In the little house he built for his parents, on Lake Geneva, he used different colors on the walls. I think it's a gem. But sometimes, when the color comes in the form of paint, it disappears and is forgotten. Architecture is often taught in terms of form. There's a Mexican architect, Luis Barragán, who doesn't get talked about much, but he's a master of color. There's also the Italian architect Carlo Scarpa, who is very sensitive. There's a museum of medieval art in Palermo where he constructs the narrative of the exhibits almost entirely with color. Ultimately, you have to work with color in architecture. And any architect who doesn't understand that is in trouble! Color is the link I see between architecture and Balcells' work. I'll can't look at the periodic table as an architectural material because architecture works with mixed material as opposed to pure ones. Eastern cultures have always understood perceptual sensitivity, but there is a certain brutality in the West. People say, "I expose the material because people need to understand that concrete is noble, that brick is noble, etc." And that's true, in part, but then there's another hidden effort going on, because in the end you realize that there are no pure materials.

You've also thought and written a lot about the idea of the city. The city may seem like a chaotic and bustling place to us, but in the end it can be understood as a kind of structure that organizes a group of people, in the same way that the periodic table is an ordering of the elements.

When coexistence and the complexity of a society require a diversity of functions, the urban pattern emerges. Before the city, that distribution of social functions wasn't there. I like to explain the origin of the city as the moment when it takes on a pattern. I call it "the footprint." In reality, it's a sort of shadow. When society starts having priests who are set apart from scribes and artisans, merchants or princes... that's when there's a shadow that expresses that diversity. But it isn't always an order. The urban order appears much later, with Greek culture. Theirs was a non-egalitarian, non-democratic society (Greek democracy is a myth), but there was a certain equality because of a drop in the social and economic level, and the city offered an expression of that equality, and that's when the regular layout of streets emerged.

If we say that order in cities began in classical Greece, it is relatively recent: 2,000 or 3,000 years ago. But the origins of cities go hand in hand with the emergence of agriculture and sedentarism, which dates back to some 10,000 years ago. Today, a world without cities seems unthinkable to us, but artistic manifestations are much older – tens of thousands of years old. Art seems like a much more essential thing for humankind, but at the same time there is a certain separation for most people now with respect to art. Cities, on the other hand, are ubiquitous, even though they are a relatively recent invention.

Art, as we began to understand it in Romanticism, as a superior, aesthetic, pure experience – I don't think that existed before. We can have that experience when we're looking at a Greek statue, but I don't think the Greeks did. I think the statues were simulacra of a deity. The mysterious and almost unbearable beauty, which they do have, comes from our reading of them in the present. On the other hand, the origin of sedentarism also needs to be rethought. The prehistorian André Leroi-Gourhan studied prehistory in a different way. Instead of looking at what remained, he looked at what was gone and the debris it had left behind. Studying garbage, which is a kind of periodic table of the urban world, can help us understand who we are. And according to that, in Europe 40,000 years ago, villages were places of repeated occupation. Sometimes for as long as 10,000 years. That's much longer than the area of Barcelona has been inhabited! And that occupation persisted not only because the sites were suitable, but because the connection to the sites were passed down from generation to generation. That somewhat undermines the idea that we only started making cities once we became sedentary. I would say it's the other way around. We aren't rooted in places like that now. There are cities, but we live like we're rootless: when we move to a new place, we throw things away.

> In his essay "The General Theory of Garbage", the writer Agustín Fernández Mallo says that garbage offers a lot of information, but that the digital world doesn't generate the same volume of garbage – or at least our garbage isn't as clear a reflection of how we live. Future historians will find it very hard to understand us.

Maybe so, but I don't want to think about all the plastic in the oceans now, and how much fuller they will be of phone batteries. We're generating more garbage than ever and there's a kind of hysterical mobility. If you don't travel all the time, you're nobody. Now it's coming under scrutiny because of the energy issue, but also because the pandemic has taught us that unnecessary mobility can lead to disaster. And I don't say that because of the question of being rooted, but because I think we need to better understand the benefits. Going back to a place you know and love is a fantastic experience for human beings. When Jaime Gil de Biedma talked about his house in Nava de la Asunción, he used an expression that I really like because it says so much more than what you might expect from language: "I always end up coming back." There are places we'll always go back to, and it's good for us to value that, because they make up our biography.

And I love having cities that I visit from time to time, that I'm familiar with. Although they aren't my home, I love them; I have an emotional relationship with them.

> Historically, cities have always been a physical support for art, but certain contemporary pieces require a more complex support than they once did. Many of Balcells' pieces, like *Universe* or *Frequencies*, require a very specific space. Is there a relationship between cities or urban space and the manifestations of art?

We often use the word art to refer to religious icons. There are Caravaggio paintings that hang in churches, and much of what we see in museums from the ancient world are objects associated with buildings. In that sense, I question historical museums more than contemporary ones, because I would prefer for contemporary work to be imagined for a museum rather than for a private collection or the world of speculation. Historical art museums, on the other hand, wrest objects from their natural places; they displace them. We were talking about uprooting before and now about displacement. And then the work ends up having a different reading. Velázquez painted for the ugly king, Philip IV, and the work wasn't for the public, it was for the king, who also bought paintings. Some scholars have said that contemporary museums are monstrous, that contemporary art is monstrous, but it's just the opposite. Now we've created spaces where we go to look at all kinds of non-functional, ideological and symbolic objects.

Cosmic Thinking

David Jou

The first time I went to a lecture by David Jou, he started by asking the audience a question that confused us: How long does it take for a bacterium to form? Everyone looked around, puzzled, until someone at the back of the room shouted: "700 million years!" Finally, I understood the question. The Earth was formed 4.5 billion years ago, and the first life form, a bacterium, appeared 3.8 billion years ago. Therefore, it had taken 700 million years for those tiny, pioneering beings to be formed. But Jou looked out at the audience with a sly smile and said, "It takes 10 billion years." What? He went on to explain: in order for the Earth to be formed and for it to contain the chemical elements necessary to make bacteria, first there had to be a generation of stars that had to be born, and evolve, and create those elements by way of thermonuclear reactions. Then there had to be enough time for those stars to die and for all those new elements to spread around the universe. Because the universe is about 13.8 billion years old, the Earth was formed 4.5 billion years ago, and bacteria took 700 million years to appear, it actually works out to 10 about billion years. This broader vision of things, this deeper perspective, is no doubt what led David Jou to pursue physics, poetry, and theology. As we can tell from his words, he is someone who delves into every layer of reality with the idea of scratching away at its deepest nature, its most impenetrable depths. Ultimately, a cosmic thinker.

Physics, poetry and theology form an interesting –
but unusual – triangle.

I started publishing poetry when I was 15, two years before I started studying physics. I haven't stopped since, so poetry has always been an important part of my life. At the same time, I've often been asked to talk about questions of science and faith. Although they may be very different things, looking into what kind of dialogue can take place between those two languages has captured my interest more and more. That triangle formed by poetry – which is an aesthetic dialogue with the world, theology – a metaphysical dialogue with the world, and physics – a dialogue with the world, is incredibly stimulating. And all that led me to try to express, in a different way, the existential, emotional, or rational surprise that comes from certain results or certain questions in physics. And I've dedicated whole books to that, books of essays or popular science. Sometimes, although it's possible to express those things, we're still left asking ourselves what exactly it is that – as a person, not as a specialist – we should find moving or surprising about the answers that come to us from physics. And I've looked for an response to that through poetry.

How did you find out about *Homage to the Elements*?
I found out about the piece through the periodic table. I wrote a poem about the periodic table, and Eugènia Balcells asked me if she could use it in an exhibition. That was the first time I had heard of her, and I started exploring her work from there. I thought her presentation of the periodic table in *Homage to the Elements*, with the series of colors in the visible spectra, was extraordinarily beautiful. In the Physics department, you're introduced to three or four of the known spectra. Having put them one next to the other in a series of fine strings of different colors, having arranged them in a series of fine strings of different colors, has an incredible emotional and aesthetic power. It surprised me, and it made a strong impression. And I was disappointed it hadn't occurred to me before.

Since then, I've looked further into her work, with *Frequencies* and *Universe*, her work on planets and planetary music, the spectrum of vibrations that our ears can't capture. Translating all those inaudible sounds into frequencies that we can hear and being transported by the mystery of those sounds reminds me of whale song. Listening to the sounds of whales communicating at the bottom of the ocean is like seeing a new reality. In that sense, I've found those three pieces by Balcells to be both fascinating and riveting.

How can Eugènia Balcells' work be interpreted from a theological perspective?

One of the ongoing problems in the world of religion today is how to express religious ideas symbolically. There were once very successful representations, which held true until the model of the world began to shift. Now, with a different world model, we need to find new ways of expressing a series of spiritual and conceptual problems from the aesthetic realm. It hadn't occurred to me before, but drawing on Balcells' work, if we wanted to build a chapel, for example, and we put a selection of the spectra of some of the elements into the windows, it would create effects which – without referencing God directly, but rather referring to the work of God or the richness of the world – would generate a completely different atmosphere from the one created by classic stained glass windows, whether figurative or abstract.

Universe is also a kind of extrapolation of *Frequencies* because the planets are translated into audible frequencies. Plus, it has the added appeal of expressing it aesthetically and in movement with the screens that rotate and generate different aspects of each planet. As opposed to being a fixed positive image, like in a Power-point presentation, *Universe* is full of life, where there is the sound on the one hand and the movement on the other. From a theological point of view, I could imagine a spinning cross, with Christ on one side and nature on the other. As it turned, it might offer different images to evoke the feeling that behind the pain, behind the injustice of the world, there is something that offers redemption. We could call it "Hope".

Ultimately, Balcells explores a series of aesthetic resources which, if you let yourself be transported by her creativity, can later be harnessed in different ways. There are some pieces that you think are interesting when you look at them, and every time you see them there is a pleasure in it. But there are others that go beyond that because they inspire you and they get under your skin creatively; they invite you to explore further. Balcells' reflections and research are of a similar kind: they are suggestive, and they invite you to follow them in your own way, freely following your own path. They offer resources for investigating questions that aren't exactly the ones she works on, and they open up avenues for exploration, like the case of religious art that we were talking about before, for example.

> *Homage to the Elements* also reflects a certain idea
> of creation: it is made entirely out of the elements
> from the periodic table, which we can identify
> by their unique light. That ties in with scientific
> cosmology and also with theology. How do those
> two disciplines relate to one another?

Physical cosmology and theological cosmology ask different questions and they aren't the same, but one can inspire the other. Physical cosmology deals with the fundamental laws of the universe, which elemental particles make it up, how the structures, both large and small, emerged, and how and when the universe began. In contrast, theology deals not so much with the moment when it began but, above all, with the meaning of the universe, the mysteries and the surprise of its existence. What is strange is that the existence of a universe like ours should be surprising for physics, because life can only exist in a universe with specific values for the fundamental physical constants. It's almost a metaphysical surprise: why does this universe have these laws, these physical constants that provide for the existence of carbon? Based on those considerations, we can establish a series of dialogues, always from the standpoint of surprise, as opposed to assurance.

> Throughout your life, has your deep faith in
> something beyond physics evolved?

Not much, especially because, from early on, I was drawn to the idea that love was the foundation and the destiny of the cosmos. When I was little and I was told that everything is fueled by love – which is what Dante and Saint Thomas Aquinas said – it amazed me. In the end, how things work is less relevant than the purely hypothetical fact that the most important thing, at the base and at the pinnacle, at the foundation, at the origin and the endpoint, is truly love. It's a marvelous idea, which has always inspired me and interested me. In that sense, the intense emotion that I felt when I was a boy hasn't changed, but I have become more aware of certain rational objections.

> What is the role of technology in a Western society like ours that is increasingly materialistic and is less interested in questions like those?

Society evolves, and so does theology. In fact, theology is always asking questions. It's a constant questioning. Just like physics studies life, storms, the oceans, or molecules and relates them to a more abstract and deeper reality – that is, the fundamental laws of the universe – theology is continually asking about the different types of society, how they represent freedom, responsibility, social organization, death, the origin of life, or the fate of things. And it doesn't relate that with things in themselves, but rather with a deeper and more general principle. In that sense, anyone who knows a little about theological problems can take an interest in them, even if they aren't believers.

On the other hand, materialism should be an almost mystical surprise. Just looking at the complexity, the subtlety, the relations between all the molecules in a single cell, protein folding, supermolecules, how molecules communicate with one another and recognize one another... Without getting into the fundamental questions, just thinking about the molecules in a single cell, how they recognize each other and how they regulate their rhythms of chemical action, is an absolute wonder. For me, materialism is almost a form of mysticism. I've read the ancient materialists, Greeks and Romans, and materialism as a theory doesn't bother me. It's possible that everything is material, but it's a possibility that physics doesn't account for. According to physics, at the origin of the universe there was both matter and anti-matter, and they should have annihilated one another and become light. If there hadn't been a break in the symmetry between matter and antimatter, the entire universe would be light. There wouldn't be a single atom, or a single particle of matter or antimatter. In that sense, the presence of matter in the universe today is an extraordinary existential surprise.

Following this shift from matter toward light, light is
one of the fundamental elements in all Balcells' work.
It also has a relevant role in theology, doesn't it?

White light had been used as a metaphor for God until Newton. It
was considered the purest existence, pure understanding. It wasn't
made up of parts, and what gave it color was matter, in its variety. But
Newton observed that light split into colors when it passed through
water or a prism and when it formed a rainbow, and he wondered
how it was possible that different materials always dispersed light into
the same colors and in the same order. At that point, he asserted that
white light is the sum of different colors. And that caused a kind of
surprise in the theological metaphor, which was purely metaphorical.
Now one of the interesting things about light is background radiation,
microwave radiation, which is the result of the annihilation of matter
and anti-matter at the origin of the universe. So, each of our electrons
is a survivor of 200 million protons and 200 million antiprotons that
were annihilated and gave rise to 200 million photons. It's incredible
to think that every particle that makes us up is one survivor of the 200
million photons that surround us.

The emission spectra on which *Homage to the
Elements* is based can be explained through quantum
physics. In that sense, the piece also uses that theory
as a tool to generate profound understanding of
the world. What kind of dialogue is there between
quantum physics and theology?

It's a really interesting dialogue, and not just from the point of view
of Christian theology, but also Eastern theologies and Kabbalistic
theology. Quantum physics makes some surprising predictions, and at
the same time it leads to some curious questions about the world. In
classical physics, for example, something is either a particle or a wave;
if it's a particle, it's always a particle; if it's a wave, it's always a wave.
But according to quantum physics, particles and waves aren't mutually
exclusive. In keeping with that theory, we don't know what things are
in themselves, but we know that under certain conditions they behave
like particles and under certain other conditions they behave like
waves. Plus, a relationship can be established between their behavior

as waves and their behavior as particles. That shows us a reality which classical physics believed to correspond directly to the senses, but which quantum physics describes as more mysterious, more surprising: what we perceive as a contradiction (it is either a particle or a wave) comes from something deeper which, depending on its manifestations, will either be a particle or a wave.

> That theory also states that we can't understand
> things to the degree of precision that we might like.

The uncertainty principle breaks with the determinism of classical physics and results in a world that is open to surprise. The theory of evolution did something similar: on another level of change, it tells us that our world is open. It isn't a world closed off by determinism, but rather open to surprise, which can manifest in the presence of new species – like, for example, the new coronavirus that has caused a pandemic. That's very logical from an evolutionary point of view: there's a mutation in a certain virus that develops its aggressivity and transmission capability, and which changes the world. Like the Kabbalist's would say, a change in one letter in a specific word – because the genome is like a series of long sentences – can change the world.

Quantum physics brings us face to face with a reading of nature that uses a language that is beyond us, beyond the abilities of our brains. It tells us that the logic of the world, the deeper logic of nature can be very different from the logic that we work with. That sounds like a theological problem: What is the language of God? What is God's logic? Is it ours or isn't it?

> Based on that logic, the properties of a system aren't
> defined until we have measured it.

We don't know if things will behave as particles or as waves. We know that if we do one kind of experiment, they will behave like particles, and in another they will behave like waves. But so long as we aren't doing an experiment, that thing, that entity – which we can refer to as light or as an electron – we can't know exactly what it is. We can only know as we're doing the experiment. And that's also interesting theologically, because when you talk about God, you don't really know what you're talking about. You begin to have a notion of it as you're doing the experiment, which may be in the form of introspective silence or active solidarity. There can be different experiences of God. Reality, like God, is mysterious, and it can only be defined – albeit partially, never completely – through experience.

Have you had that kind of experience in your personal life?

Of course. Anyone who is interested in these things can have that kind of experience, during moments of concentration, introspection, prayer or study. Or when you're helping others, which is a moment of action, or in the moment of aesthetic contemplation. Because we talk about silence, but sometimes the path to God, or to a higher reality if you don't want to call it God, can be, for example, the experience of music – an aesthetic experience, in other words. Those three grand experiences – reason, action, and aesthetics – are part of the world of science and the world of religion, although they are read in different ways. In other words, the deeper reality to which science refers is the reality that can be observed or deduced, which ultimately ends in mathematics. In religion, that reality is God. And those two realities, math and God, are mysterious. They aren't material; they exist outside space and time. They are relations rather than essences. And that is what the mystery of the trinity says, which ultimately refers to something quite simple: God is not an essence, but a relation. And if God is seen as a relation, it's no longer seen as in opposition to math, precisely because mathematics is a relation. And love is also a relation.

Is that idea of God as a relation equivalent to the idea of God as the creator?

God as the creator doesn't mean that God is at the origin of the universe. It means something more important: that the existence of the universe depends on God. For example, like I was saying before, in physics there is always a mathematical dialogue with the world. And that leads us to wonder what kind of existence mathematics has. Is its existence outside space and time? How does abstract mathematics govern concrete physics, matter? Those abstract regularities in mathematics were around long before the existence of any planet, or any living species.

> In that sense, the mathematician Marcus du Sautoy
> says that, since mathematics has that property of
> being situated outside time and space, it can be
> considered an agent in the creation of the universe.

Absolutely. An agent that also sets limits when it comes to creating the universe. Because many different universes can be conceived. If we take the same physical laws that we know now but with different values for the fundamental constants, it will result in completely different universes. String theory states that there are something like ten to the five hundredth sets of consistent values for the fundamental constants of nature. If each of these possibilities were assigned a universe, there would be ten to the five hundredth possible universes, each different from the next, most of them very unstable, very short-lived, very small, or which would expand and then directly collapse again.

> Does Balcells' work suggest anything to you in the
> sense of a creator God?

There are different ways of imagining the mathematical aspect of a creator God. One of those aspects is to imagine Him as a mathematician-musician – rationality and emotion at the same time, intellectual and sensorial beauty at the same time. In a way, that's how Pythagoras must have thought about it, when he refers to the music of the spheres, even if he wasn't thinking of the idea of Creation. Frequencies play an important role in music, but also in quantum physics and superstring theories. They're frequencies that talk about sums, constructive and destructive interferences, plural presences at each point in space, without one excluding the presence of the others, unlike what happens with matter. What is fantastic about *Homage to the Elements* is that it represents the absolute sobriety of those frequencies, the colors of the spectra, and through the different elements, it includes nearly all the colors of the optical spectrum. If you're a little contemplative and a little sensitive, *Homage to the Elements* can put you into contact with a different reality. Whether you find God or not through that reality, that depends on you.

The Periodic Table

David Jou

Look at them: on the right, the noble gases – in red, like Sundays, like holidays, because they refuse to mix together and they're laid back and relaxed;

at the very top, like two isolated towers, hydrogen and helium, the most dominant elements in the content of the Universe – it might have been more logical to situate them as roots rather than domes, since that's what they are: origins, foundations, celestial roots;

beneath them, six more stories, and, like two basement levels, the lanthanides and actinides;

on the sixth floor, the offices of life – carbon, nitrogen, and oxygen, in all their fertility: forests and atmospheres, buried energy;

on the fifth, as we continue our descent, the sands of all the beaches and deserts – silicon, and the salts of all the seas – chlorine, sodium and magnesium;

on the fourth floor, calcium and potassium – which, along with the sodium from the fifth, flow through our nerves like dreams – and, like an impassable gate, iron.

From there, everything was formed through violence, in large explosions of supernovas: the copper from the fourth floor, the silver from the third and, from the second, the gold and the mercury – so fascinating – and the lead and the barium, so dense.

Radium – and uranium in the basement – radioactive, as though reminding us of the deafening tumult at its origins.

In the last underground level, what's there is mainly artifice; the atoms are very short-lived, a play of ingenuity that lasts just long enough to be given a name and then falls apart – they aren't needed anymore: they are a load that the world can't quite figure out how to bear.

Look at them: here, the building blocks of the world, arranged in stories, on shelves, regularly repeating properties, revealing a deeper structure,

no longer eternal and immutable matter, but history in the stars, traces of temptations, buildings with levels and sublevels, clouds of uncertainty, combinatorial flowers.

We come from beyond all these pieces, who knows where we're heading, but what a joy to have been able to discern behind them the beauty of a logic in the world!

Homage to the Elements, Eugènia Balcells, 2009

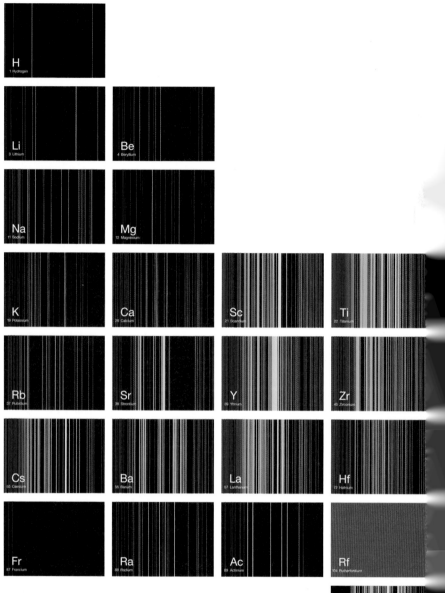

V	Cr	Mn	Fe	Co
23 Vanadium	24 Chromium	25 Manganese	26 Iron	27 Cobalt
Nb	Mo	Tc	Ru	Rh
41 Niobium	42 Molybdenum	43 Technetium	44 Ruthenium	45 Rhodium
Ta	W	Re	Os	Ir
73 Tantalum	74 Tungsten	75 Rhenium	76 Osmium	77 Iridium
Db	Sg	Bh	Hs	Mt
105 Dubnium	106 Seaborgium	107 Bohrium	108 Hassium	109 Meitnerium
Pr	Nd	Pm	Sm	Eu
59 Praseodymium	60 Neodymium	61 Promethium	62 Samarium	63 Europium
Pa	U	Np	Pu	Am
91 Protactinium	92 Uranium	93 Neptunium	94 Plutonium	95 Americium

Light is the voice of matter.

When it is separated into the colors that make it up, it reveals signatures that tell us the identities of the elements that emitted it. Each element gives off unique colors that form its light footprint. Thanks to spectrology, begun by Isaac Newton in a dark room in the 17th century, astronomers can determine which elements make up an object that has emitted light. That is how we know what stars and nebulae, galaxies, and supernovas are made of.

In the mural *Homage to the Elements*, Eugènia Balcells pays homage to the elements of the periodic table, the fundamental building blocks that time and chance have used to form life. And she does it by showing us each element's name in the language of light. A name made of the purest colors, a light footprint, or a voice that is nothing more than the color palette with which the world has been painted.

DAZZLING!
COLOR AND LIGHT

From Barcelona, we're bringing you the sounds and spectacular
moving images created by Eugènia Balcells. Tonight, together
with the chemist Roald Hoffmann, they'll be undertaking
an artistic and scientific exploration of color and light. The
work of multimedia artist Eugènia Balcells is nothing more,
and nothing less, than an invocation of the foundations
of the universe and of life: matter and energy, light, color
and the chemical elements. You'll be captivated by a visual
journey through some of her installations and films: including
Frequencies and *Color Wheel*. In the former, the light signature
emitted by each element is mixed with the others', a metaphor
for beginnings. The latter is a kind of Borgesian Aleph, a
universal catalog of moving images organized according to the
structure of the light spectrum. And Herr Möbius might just
make an appearance.

When it is separated into the colors that make it up, it reveals signatures that tell us the identities of the elements that emitted it. Each element gives off unique colors that form its light footprint. Thanks to spectrology, begun by Isaac Newton in a dark room in the 17th century, astronomers can determine which elements make up an object that has emitted light. That is how we know what stars and nebulae, galaxies, and supernovas are made of.

In the mural *Homage to the Elements*, Eugènia Balcells pays homage to the elements of the periodic table, the fundamental building blocks that time and chance have used to form life. And she does it by showing us each element's name in the language of light. A name made of the purest colors, a light footprint, or a voice that is nothing more than the color palette with which the world has been painted.

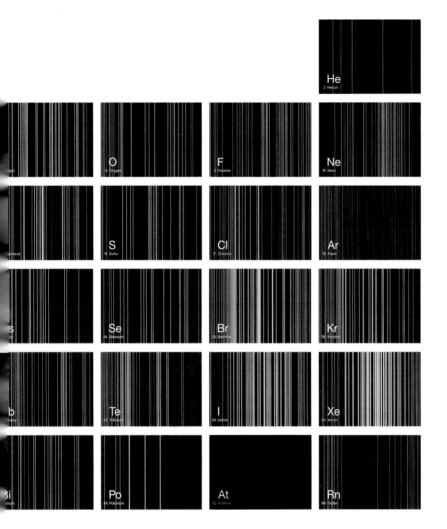
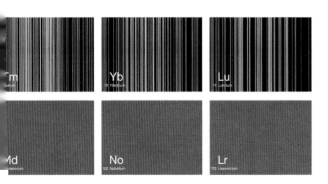

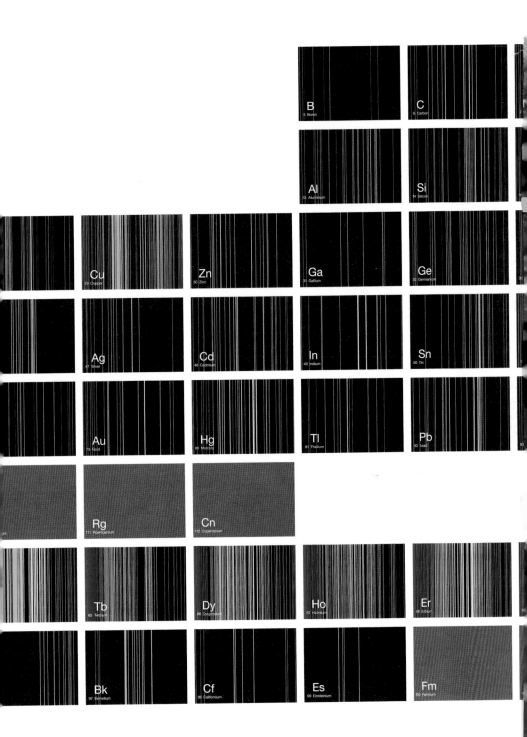

DAZZLING!
COLOR AND LIGHT

From Barcelona, we're bringing you the sounds and spectacular moving images created by Eugènia Balcells. Tonight, together with the chemist Roald Hoffmann, they'll be undertaking an artistic and scientific exploration of color and light. The work of multimedia artist Eugènia Balcells is nothing more, and nothing less, than an invocation of the foundations of the universe and of life: matter and energy, light, color and the chemical elements. You'll be captivated by a visual journey through some of her installations and films: including *Frequencies* and *Color Wheel*. In the former, the light signature emitted by each element is mixed with the others', a metaphor for beginnings. The latter is a kind of Borgesian Aleph, a universal catalog of moving images organized according to the structure of the light spectrum. And Herr Möbius might just make an appearance.

Face to Face with *Homage to the Elements.*

Roald Hoffmann

REWRIT

When God made the sun
he lay back on his white
sand beach, and reaching
out, with both pale hands,
into his space, he shaped
there a sphere of hydrogen,
God did, set it alight
with his nuclear fire. He
felt, God felt, its warmth
on his soft hand. And
it was good, it was his sun.

When God set about next
to make the moon, he put
his feet on the ice cap
of Mars, and reached out
again, seizing a piece
of an old sun, and God
threw it, like a snowball,
at his earth. The earth
rocked, and so the moon,
God's moon, came to be. He
felt its reflecting light,
and it was good, his moon.

When the time came for God
to people this blue earth,
he stood knee-deep in paddy
and sea, and, dear God, he
didn't make people in his
image, but just reached out
his now sunburnt hand
to plant a mitochondrion,
here a squid's eye, a seed
of rice. Hazard he gave them,
rules. God's time, and soon
enough the creatures came,
spoke. It was good, the word
between God and his people.

Xavier Bosch

Imagine the perfect system: an exquisite proportion of ordered chemical elements in perpetual motion, generating a biological differentiation with the ability to produce molecules, cells and organs with very different functions, through tiny modifications in the sequences of an immense list of possibilities. Tissues that support the weight of the body, organs that metabolize the raw substances, heartbeats that go on for a hundred years, neurons that stimulate and excite.

Someone has tried to put a price on a human being, calculated on the basis of the cost of the amounts of each element that makes up a human body. Six elements form the recipe for 99% of a person: oxygen, hydrogen, carbon, nitrogen, calcium and phosphorus. At market price, somewhere between 120 and 150 euros. A little more if we add the gold, silver and titanium from prosthetics, and a smidgen more again if we account for the silicon for the microchips that will connect our brains to computers and networks soon enough.

These elements, in the right proportions and balances, have brought about Beethoven's Fifth Symphony, the theory of relativity, the Taj Mahal, and space travel, as well as smallpox, AIDS, and COVID-19...

The best things in life are free.

So where is the difference between the humility of the elements that make up a human body, a sprig of thyme, or the lava from a volcano, and the greatness of the results of their infinite combinations?

Amarjit Chandan

GRAMMAR OF BEING

God uttered a word: *Kun* * Be — and it was.

Be is the seed of all verbs.
So many verbs make a noun.
The noun is the fruit.
       The seed ripened waiting to be sown again.

It's the end of the continuous verb
       happening in the present perfect
              all the time.
The noun is the ultimate verb
       of conception   of bearing that reaches
its end
         of no end.
The pyramid rests on the triangled tense.
The noun is the point on the pyramid top
       where the journey starts towards the timeless
         in the present.

* *Kun* (كن) is an Arabic word for the act of manifesting, existing or being
and consists of the letters kaph and nun. *Kun fa-yakūnu* has its reference
in the Qur'an cited as a symbol or sign of God's mystical creative power.
The verse is from the Quranic chapter, *Surah Yasin.* – *Wikipedia*

Carlota Subirós Bosch

An empty space that is tinted deeply by the rhythmic sequences of color fields, amid the overlapping reverberations of metal bells. Bodies that become silhouettes, figures that are charged with energy in the dark and the movement.

On the scene, the space and bodies are our tangible materials, the clay we sculpt with. Light and sound are our intangible materials, ethereal but omnipresent, essential, specific, decisive.

Light and sound are transmitted by reverberation; they invade the space and permeate our senses. Light and sound *animate* the space. They move it, unfold it, hide it, designate it, modify it. Light and sound give form to the passage of time and make the space breathe, like a breeze.

In the radiant nakedness of the color and timbre of the installation *Frequencies*, we feel an absolute perceptive reality opening up, as essential as it is mysterious.

Talking about the admiration brought on by the existence of something, Richard Sennet points out that a possible definition for the ancient Greek term *aura* might be "bathed in its own light".

Thus, the aura is not only the halo of light that a body or object gives off, but also the transformative effect of that light on the body itself, bathing the object, and inevitably appearing around it.

Walter Benjamin used the word to refer to the halo of value that surrounds each original, unique, exceptional work of art.

In her meditative *Homage to the Elements*, Eugènia Balcells discovers the light spectrum characteristic of each element and fixes it in a chromatic image that she herself calls their *voice*. A singular vibration revealed from the heart of copper, oxygen, iron, zinc, uranium, cadmium. An aura of meaning, a message that reaches out to us from the very origins of the universe.

In the reverberating space where the spectra of all the elements intersect, we appear. In the reverberating space where the spectra of all the elements intersect, life pulsates, an infinite miracle.

Jean-Louis Froment

PRISMA

The colors unfurl across the barcode of the invisible
placed in discontinuous waves they sink into a temporality that is familiar and
yet far off
We recognize
but
we don't know
It's the paradox of the artist's gaze
what we see is not what is known
the origin never took place
time is allogenous
The words flicker and breach the object being looked at
accompanying
abandoning
returning
The alphabet is mutilated
art
manufactures it an unnamed present
Taken at the atomic word of the images
the radiations of memory step in
the universal falls into rapture and shines unaware
The memorial event is inevitable
like a refuge weak from being looked at
We feel
like a long pictorial fainting
Rothko
like a rational construction of color
Albers
Perhaps
the passing of a solid breath of science fiction
hatched/striped
Surely
a primitive halo
elemental chemical artifice of diabolical emissions
Spectrum
Fear
Atom

Then from the windows that open onto the garden, in the prism of my gaze, I bring in the colorful gradations from the winged angel by the miniaturists of the Italian Renaissance, when the heaviness of the pigments keeps him on the ground for an instant, as I imprison him in the words that are mine.

Messenger of lightning, atomic intermediary, the angel with the colored wings is made to be perceived, only perceived, sometimes in the radiation of a rainbow, sometimes in the imperceptible filaments of light, sometimes in its imaginary electromagnetic flights which are, with us, on their way toward the sun.

Salimata Wade

At first glance, *Homage to the Elements* by Eugènia Balcells – the mural created in 2009 which, 10 years later, in 2019, became the emblem of the International Year of the Periodic Table – seemed like an example of what a work of art can be: a simple and effective tool for cultural and scientific mediation. It is an original, innovative and very widely shareable production. I see the mural as a powerful, and at the same time playful, means of uncovering lots of people's interest in science, regardless of the place on Earth where the work is seen.

In effect, *Homage to the Elements* is a kind of alphabet of science, which speaks directly to our senses. We perceive its kaleidoscope of colors immediately. This means that, in addition to the precise significance it has for chemists and physicists, the mural's chromaticism also raises many questions for the everyday people observing it.

I am a Senegalese woman who, as a university professor, is interested in the geography of African cities, and, as the creator of the Compagnie du Bien Manger, I work to promote good nutrition among citizens who are increasingly moving away from local products and their ways of cooking and consuming them. From this double perspective, I believe that *Homage to the Elements* provides an extraordinary possibility for people to understand the connections between human knowledge and their daily lives.

Recognizing the colors of food and knowing that it carries nutritional qualities related to the primordial elements is as important as understanding the landscapes of our lives and knowing how to adapt to them in a sustainable way.

On the other hand, raising the question about the relationship between modern science and traditional African knowledge offers a very broad educational dimension, of which *Homage to the Elements* is an excellent example.

I would like our artists to feel drawn in by this perspective and to convey to our young people the vision humanity now has of itself thanks to our search for the unknown. And because chemistry, for example, has allowed us to transform the food industry so that it can meet the needs of a growing human population, at the same time we must have the necessary information to understand its limits – all too often surpassed – in order to preserve the food cultures that have demonstrated their benefits over the centuries.

We can find many examples of how the sciences influence the processes of urbanization and globalization, but without the necessary mediation, those questions fall beyond the understanding of a very significant part of the population. When that happens, there can be no possibility for a critical spirit, and the continuous learning that we all need to coexist in community becomes very difficult.

I have no doubt that the arts are unquestionable sources of tools to expand humanity's vision beyond what is urgently immediate. And I also celebrate that it is a woman who is revealing to us the innate beauty of the elements that make up matter and the responsibility of keeping them in balance, as required to sustain life.

Jorge Ventocilla

I remember it was in mid-July 2017 when I attended the opening of the exhibition *Light Years*, at the Museum of Contemporary Art in Panama City. "Jorge, you should be there," Alexandra Schjelderup, the director of International Cooperation at the Ministry of Culture, insisted when she sent me the invitation.

When I arrived, Eugènia was already addressing the crowd. The room was packed and from the back corner, where I was able to find a space after a long series of "Excuse me"s. I listened as she talked about her mural *Homage to the Elements*,

– "Each chemical element of matter has a characteristic light spectrum: it has its own voice…"

– "The elements are the alphabet of life, and they make up all of reality by dancing together…"

I heard her words as though they were dedicated to me personally.

"Light is the voice of matter" […] "Thank you. Thank you to all the elements," Eugènia Balcells said at the end of her speech.

I studied biology (zoology), and for more than 30 years I worked in environmental education and scientific interpretation, as part of an organization that is well known for its scientific work in tropical biology. From very early on in my education I sensed that science, art and even spirituality come from the same Big Bang; and that, inescapably, they will keep converging, getting closer and closer. My work as an educator and communicator of science has advanced along those lines.

But I haven't found too many people capable of bringing together those three branches of human knowledge – science, art and spirituality – and honoring their union. I saw it that night at Eugènia's exhibition. And I want to say now, happily, that it encouraged my style of working.

I had to wait a long time to be able to talk with the artist. There were a lot of people there, and everyone wanted her attention. Plus, the exhibition and its message had made everyone talkative. Suddenly I saw an opportunity. I approached Balcells and whispered in her ear: "Your mural can help re-enchant the world. Thank you for helping us to understand." I remember that she looked at me and flashes of complicity passed through her eyes. I walked away happy.

Maria Nadotti

KNEADED WITH LIGHT

Dear Eulàlia,

It is to you by the gentle name ("she who speaks well", "who speaks softly") that I write these brief thoughts that came to light by looking at the works of Eugènia Balcells, *Light Years, Universe, Frequences.* She too, like John Berger, our mutual beloved friend, looks and wonders, looks again, literally respects (the magnificent *re-spicere*, looking back, of the Latin language) what her eye rests on and does not necessarily see.

There are visible things that the human eye does not see and others, invisible, to which that same eye is blind, but that we can perceive with our other precious senses – the tenderness of the hands, the intensity of listening, the nose, the taste buds. Between the colors that the light reveals and the sense of taste there is a sort of chemistry: there are paintings that one would like to taste, to lick, and others that one would smell, like cats do with what attracts or repels them. I think of the frescoes by Piero della Francesca, so logical and so mysterious, where carnality emerges as an indisputable theorem from the backgrounds that form the backdrop to his narrations. I think of the paintings by Hieronymus Bosch, which smell of sulfur and myrrh, of death and a tenaciously vital entanglement.

Eugènia Balcells with alchemical, generative intuition goes to the origin of the scopic drive: matter is light and gives light. We are irresistibly attracted to the complex, multiple, never repetitive combinatorial dance of the elements that make it up and break it down.

At the moment of birth, of our "coming to light", well before things are distinguishable from each other and we from them, long before we are able to recognize them and associate them with a name, we humans float on an indistinct continuum, in a joyful and terrifying seriality. We are matter in matter, we make and receive light. Then, little by little, contours, borders and darkness take over. We identify and identify the other from us. Seeing, then, can coincide with blackness.

Jaume Casals

When I was a teenager, I thought I would be a scientist.
Switching to philosophy, which I don't think is properly one
of the arts but has ended up on the side of things that are not
experimental sciences, was a last-minute choice. I'm a bit of a
fetishist, and at the time there were certain things I really liked,
like the periodic table. I even bought some plastic flash cards
with the chemical elements on them – and the declensions of
Latin verbs. Plus, my father was a chemist, and he had a little
lab at home. In his lab, I'd use the Bunsen burner, and I'd heat
Erlenmeyer flasks using wire gauze with a circle of asbestos
in it, and every now and then something would blow up in my
face, or I'd get silver nitrate all over me, which is dangerous if it
gets in your eyes. I poisoned myself a lot in that lab! I did some
pretty stupid things because of my chemistry fetish. With all that
history behind me, when I first saw *Homage to the Elements*,
I had a feeling that was like looking at those flash cards or a
slide rule, which is a small, sophisticated, and very powerful tool
because it offers a lot of information in a very concise way. I had
the feeling of being in the presence of a great work, something
monumental. And it occurred to me that the piece, which almost
isn't a work of art but an idea, had been turned into a monument,
that it involved some kind of miniaturism in a more mental than
physical sense – in other words, something minimalistic and
profound at the same time.

Holly Fairbank

HOMAGE TO THE ELEMENTS:
ENCOUNTER WITH NEW FREQUENCIES

When I first encountered *Homage to the Elements* in
Eulàlia Bosch and Eugènia Balcells' studio in 2018 in
Williamsburg, NYC I was thrown into an astounding
sensation, or perhaps emerging awareness, that the mural
before me was altering my understanding of how to view
a scientific illustration of the elements of the universe I
had previously taken for granted. That there was another
profound and joyous, enlightened way one could conceive
of and experience the "explanation" of these, for me, hard
to imagine elements of the universe. The very periodic
table itself, trustworthy and practical, was reintroduced
to me as a shimmering, colorful, newly identified force
of life – opening me up to new ways of perceiving. I was
reminded of a quote by Maxine Greene, "I still wonder at
how unaware I was of so many frequencies." These elements
of the universe were speaking to us through their unique
compounds and frequencies, their unique places in the
cosmos. Eugènia had tapped into each element's individual
and collective frequency and discovered a way for them
to be experienced by us through our senses, our ability to
make meaning, our aesthetic sensibilities, and through our
human gift of metaphor.

As is stated in the description of *Homage To the Elements*, it "recognizes the symbolic character of the periodic table by adding the light signature of each element to its chemical description, proclaiming that light is the voice of matter." The VOICE OF MATTER! Engaging with this work of art was, for me, to be moved to "see things as if they could be otherwise".* I was immediately prompted to reach out to Steve Noonan, the principal of The Maxine Greene High School (MGHS), and urge him to come and see the work. Eugènia believes this work belongs in public spaces, places of learning, places where transformative thinking can take place. Within a year *Homage to the Elements* was installed in the lobby of the Martin Luther King, Jr. High School campus in Manhattan where The MGHS is housed.

Homage to the Elements is as much a revelation of science in relation to art as it is, perhaps more importantly, a testament to the possibilities of the human imagination to see what is not seen yet, those things that sing, dance and can transform us. This dazzling work of art brings into being new ways of perceiving our world, recalibrates our relationship to the elements of the universe and even asks us to reconsider the value we each bring to this extraordinary conversation.

Irene Martin

REFLECTIONS ON *HOMAGE TO THE ELEMENTS*

After almost forty years of being involved in the art world organizing, producing and managing art exhibitions of all kinds, I am familiar with the role technology plays in the production of art. However, it was only when I met Eugènia Balcells that I encountered artwork produced from the convergence of science and technology with such brilliance and simplicity. I marveled at her ability to produce a beautiful work of art that explained the wonders of science made visual through technology.

On seeing the work *Homage to the Elements* installed for the first time, I thought it was Eugènia's powerful imagination that had produced those colored bars. I thought of them as a way of identifying the elements through made-up colorful "bar codes". Then I learned that it was more than just a beautiful creation. She had used her scientific knowledge and found the technology to be able to capture the actual emanations of light from each element; a few were blank because it was not possible to capture them through current technology. The familiar chart of the elements, with their names within squares in black and white, was now presented in the colors emanating from them. For a layperson who never did well in science, it was a revelation that those letters in black and white were really made up of colors – and therefore everything in our world is made up of those colors.

My thoughts went to the basic combinations that we learned in grade school science, that common salt is made of sodium and chloride, and water is hydrogen and oxygen, and bronze is an alloy of copper and tin. They were just combinations of words to me. But now they were colors in the place of the words, and they were real colors, however ephemeral they might have been. Those ephemeral colors were captured through technology and reproduced to create a work of art.

I was with Eugènia in Mexico City when she installed *Homage to the Elements* in an exhibition at the National Center for the Arts. It was part of the exhibition *Frequencies*. Seeing the installation on the wall in a room of its own made a huge impact. There was another installation that related to *Homage to the Elements*; it was a projection of the mixing and blending of the elements' colors onto the entire opposite wall. The projections were mesmerizing and fascinating. What I was seeing was what the elements do in actuality. Everything on Earth includes part of those elements mixed and blended.

I went back to look at the static images and was drawn into the work as a whole and to each element. I came to the realization that I was looking at the elements of creation; the beginning of creation, as well as the future. It was evolution and all that the future will bring.

I was also looking at the micro and the macro worlds; I was looking at the elements in my body and the mixing that created me. It was the miracle of creation, evolution as creation, and I was looking at the source of all that I am, a microcosm of the universe.

My reflection on *Homage to the Elements* was a meditation on creation, humility, mystery, wonder and all things beautiful, great and small.

Noni Benegas

HOMAGE TO THE ELEMENTS

> *More than the results or the works, I care about*
> *the energy of the author, which is the substance of*
> *the things he is waiting for.*
> —Paul Valéry

From the beginning, she was looking to solve the puzzle
of each thing, each entity or presence in the universe
– beginning, for example, with the Statue of Liberty
(a tribute to the concept of freedom in the form of a
monument at the entrance to the port of New York), which
Eugènia made the object of her exploration in 1986, for
the celebration of its centennial.

After searching the United States far and wide, she
recovered hundreds of postcards of its image, in all
shapes and colors, from all possible angles and in all
kinds of light.

She grouped them into categories, and then, by
contrasting the different colors, she reconstructed, card
by card, the likeness of the enormous copper and iron
sculpture – with its crowned head and torch held high –
in a six-meter-long mural.

After decades of research and discoveries, which
consolidated her career as a video artist through
unforgettable installations and pieces, in 2009 she
glimpsed, astonished, the key that could unlock the
shifting construction of the visible world.

And just like rice is planted with rice, she lucidly recognized the hidden source of that energy that sustains the universe, provided, in its turn, by light.

Accustomed to breaking down the puzzle of each object or presence piece by piece, she investigated the basic elements of their composition – oxygen in the air, sodium in salt, sulfur in eggs, neon in light bulbs, gallium in LED screens, etc. – and she saw in the spectroscope that, when excited or heated, they emitted a spectrum of light in vivid hues.

Something like a color barcode, specific to each one and invariable. Or more exactly: the personal light signature of each of those 118 elements the world is made of, and which are listed on the periodic table.

Death does not eliminate the components, it only separates them, breaks them up and recombines them into new things that change their form like the iron in our blood, which comes from the explosion of a star.

They are, then, the alphabet of life, which Eugènia put together in the six-and-a-half-meter mural *Homage to the Elements*, which inherits – of course! – and at the same time resolves the enigma hidden in her early six-meter mural *Liberty, A Symbolic Puzzle*.

Santiago Álvarez

SUNSET

From Vienna to Barcelona
fleeing from darkness we
fly towards the sunset.
To the right the gilded

sky bleeds through
clefts in the clouds,
to the left the shadows.

Unfair race
won by the night
stitching the wounds
through which the sun
was bleeding.

Thane Lund

Eugènia Balcells' artwork arrived into my life at the same time as Eugènia, the person, arrived as a friend. The magnitude of wonder and fascination that animates her artwork is inseparable in my experience from the passion she exhibits towards this mysterious gift we call life. Our discussions over the carpentry projects I was assisting with became a prelude to the expansive power expressed in her artwork.

It would not take long for a spark to make fire as our conversations shifted from the details of the job at hand to a macro view of our existence on this wonderfully abundant yet fragile planet. Although Eugènia and I are generations apart in our mortal bodies, we are but children in comparison to the eons of time that have been orchestrating the universe. It is this youthfulness of spirit that I have found most captivating. With a childlike wonder she has translated scientific discovery to a language we can better observe as our own.

I had the privilege of seeing *Frequencies* projected for the first time not in a museum but in an exhibition that was coordinated in Eugenia's studio in Brooklyn, NY. The work itself carried a personal tone seeing it presented this way, in private and with Eugènia present. As the colors in each element's light spectrum illuminated the walls and undulated through the air, Eugènia was there, radiant in the beams of colored light, smiling and dancing.

It is this spirit of wonder that I hope others can experience when they view *Frequencies* and *Homage to the Elements*. There is a dance that precedes our lifetimes, which has given form to the universe and every atom that we know. It is a spectacular opportunity to be part of this ongoing movement that began in the cauldrons of the stars. May we, in our time, better understand and appreciate the star matter from which we are made.

Afterword

A Journey through Light
Light

Rosa Olivares

Light is what lets us see. The darkness of a world without shadows, or rays of light, or bolts of lightning is the return to chaos, the end of humanity. In the beginning there was light, and – it seems – this is the first text, the first words in the book that bears the title *My Name Is Universe*, which at first glance might seem like an exaggeration if we were only talking about an artist. But that isn't the case.

Light (from the Latin *lux, lucis*) is the part of electromagnetic radiation that can be perceived by the human eye. But that's just the beginning of a description that can end up turning into something incomprehensible. For us, light simply defines the colors and separates day from night, the known from the unknown, the visible from the invisible. And, thus, by extension, aside from our visual ability, understanding is also considered "light". The understanding of truth, the clarification of mysteries. "A light came on", "I saw the light", now I understand. In cartoons, comic books and illustrations, when someone finally understands something, or when they have a brilliant idea, it's represented by a light bulb, one of the symbols that development has associated iconically with light. When a situation or a mystery is elucidated (pay attention to the verbs here, they're all related to illumination and luminosity), we talk about *shedding light on it* or new information *coming to light*. By extension, light is not only clarity, brilliance, that original glow, but knowledge, wisdom, truth.

God, Jehovah, Yahweh, and His angels and archangels always appear in a ray of light, like certain superheroes, and that same ray is the weapon the gods wield against evil. When Paul fell from his horse and converted to the true faith, the cause was a blinding ray of light. In *Blindness* (1995) José Saramago writes about a world in which everyone, or almost everyone, has suddenly gone blind. No one knows why or what the possible cure may be. It's a metaphor like the ones he uses in almost all his stories. What would happen if most of humanity suddenly lost the ability to see? What if the lights suddenly went out? When the power goes out in your house you bump into the furniture, the walls; you can't find the doors, even if you've been living there your whole life. We would all be lost without light, at the mercy of those few who can still see. We would be truly lost, unprepared to move around or communicate. It would be ghastly; it would be the cause of endless abuse.

At the bottom of the sea, in the deepest trenches, there is no sunlight. It is always night, and darkness defines everything. The beings that live there don't have eyes because they don't need them. They have other tools to move around: no doubt a highly developed sensitivity to sound, and magnetic fields, that keeps them from colliding with everything, like an ultra-sensitive radar; it guides them through a different world, terrifying for all of us.

Humanity has always been enlightened by the sky, the sun, the moon and the stars. Later it was fire, in bonfires, torches, and with science, electric light; knowledge keeps illuminating us more and more... although it isn't always too reliable. However, humans lived through a millennium full of monsters, fears and indefinite threats that lurked in the dark and only came out at night. The part that shares all our time with day. The monsters come out at night, with the darkness. Fear came from the night, harboring all the monsters we can imagine. That faceless danger that we know is hiding under the bed, or in the closet of our childhood room. We're all frightened children when we turn off the lights at night and are left alone in the darkness. Even noises sound different in the dark.

I talk about darkness to define the light, like the hollowing out of a sculpture, like the shape of your body in our empty bed. Darkness is the terrible part of light, its absence. And men and women have always fought so that light – knowledge, wisdom, truth and justice – can reign in this world.

Light is at the origins of art. It is light that lets us see, and therefore it is light that creates colors. It's at the origins of all the arts, but especially the visual arts. People's tastes "come in all colors", but without light there would be no colors, there would be no tastes, there would be no painting, no photography, no film. Without light, Eugènia Balcells would have had to be a scientist, maybe an astronaut, maybe a witch, a healer, a shaman. Without light, Balcells would never have been able to base her first pieces on black and white, like the play between pairs of opposites: day and night, light and dark, white and black, women and men. Along the way she would find the grey, the tones, the sounds and their own colors, the music... and today we can say that Eugènia Balcells is a little of all that and all this, since her work exists between art and science, between light and semi-darkness, between the natural and the artificial. And, more than an artist, after so much time, so much life, today she is an intermediate being who has the qualities of a scientist and a sage, a magician and a witch.

Science and magic are our two ancient forms of knowledge, our ways of transforming our reality. That communion between human generations and the elements gives way to art, to magic. And that territory is where Eugènia (her name already gives us an indication) Balcells works.

There are rare, incurable diseases that affect only one person among millions. They presage a cruel and early, not necessarily swift, death. These diseases aren't hereditary, and they set on suddenly, without warning, like in a genetic leap, an error in the perfect human construction. They're the odds of someone who becomes a millionaire in a game they play just once, but in reverse. There's nothing leads you to think you'll be the one among millions who will have it. The same goes for art. The same goes for magic. It's a mystery that begins as if by enchantment. The people who study art in schools or universities have got it all wrong. There, in those places where you may learn the technique or the history, you won't catch the evil eye, that isn't where artists are made – perhaps artisans, careful artists, who use their hands and their tools. They don't talk about feelings there, about pain, about life, but about ideas, techniques, history, aesthetics – in other words, they don't talk about the cause but about the consequence.

Art happens rarely, but it is collected in special places, it is stored apart from the rest of the world, in museums, temples that should be respected, where people talk softly, like in churches, in whispers, impressed by the excellence, the magic, the spell of those works, of those gestures, of those heartbeats that we can still hear inside a sculpture or behind a canvas. Perhaps music is where we can feel it the easiest. Art, when it arises, explodes with its own life, and surprisingly many try to disconnect it from that origin, from life. The birth of art, like that of those terrible diseases that kill you, that silently destroy you, is an unexpected seed, a tumor that suddenly crops up from a touch, from a memory, from simply being you, unique and different, like all those who – even if they may not know it – are unique and different.

In 1998, Eugènia Balcells said in an interview on the occasion of her exhibition at Tinglado 2 of the Port of Tarragona ("In the Heart of Things", 1998): "I'm tired of working for the art world. I don't think you need to know who Duchamp was to feel and understand art. My work now is aimed at people who have never heard of him." Perhaps in that moment she is only putting it into words, giving a voice to the final purpose behind her work. She is publicly clarifying her profound freedom, her way of constructing a dialogue with the outside that is increasingly free from obligations and permissions. She is stepping out of the art world, out of the structure of what we understand as art. A structure so rigid that it makes hesitation impossible; any gesture can be considered treason, desertion. And deserters are isolated or eliminated, two things that are almost a single thing, one and the same.

The exhibition, simple and intimate, ("In the Heart of Things") was truly a meticulous and refined effort. An installation based on manipulated and altered objects, video, photography, music, etc. in an empty port warehouse, not in a museum, not in the city center, not in an elegant neighborhood. "I think that artists are just a kind of antenna for transmitting" (Balcells in the same interview cited above). That is simply the essence of what Balcells usually does and which may seem simple to us – not because it is, but because we can get close to it easily enough. It seems easy, problem free, exempt from all the complicated verbiage that usually accompanies artwork and its staging. Because Eugènia Balcells was one of the pioneers of conceptual art in Spain, and one of the pioneers of video art, and of course appropriation, although we didn't know it then, and installation art, and the feminist discourse... and mixing forms and languages (or hybridizing, as an intellectual would say) – although that wasn't her place, or at least it wasn't the place she seemed to be heading to from the beginning. But life, or perhaps death, like two rocks that make a spark as they collide, just a fleeting glimmer, would redirect her steps, would change the direction of her path, her ideas and her feelings. Her life. A long journey back to the starting point, a closed circle that defines infinity and brings it back into the light. A long journey where the starting point and ending point are the same: her own heart.

In all her work, in all her installations, although they exude simplicity and clarity, there is the germ of something that lights them up like the light of a sun that she keeps among her rocks and shards of earthenware, seashells, reclaimed from the remnants that nature returns to us once they have been wrung dry. And in all her work, even the most entrenched in technology, the human element always appears. Magic, like light and darkness, like the thirst for knowledge, is always there, in the heart of her things.

In the beginning, there was "The End" (1978-2010), an extended compilation of images that close out a love story, the end of the narrative. A story in black and white, cut from the newsprint of photocomics and black-and-white romance movies, 100 images of the kisses that we always see before the words "The End" and the fade to black. It's where the enthralled reader closes the pages of the magazine and sighs, when we leave the theater after the film, longingr for something we've never had and yet that we feel we've already lost. We enter the fade to black fade of absence, of misunderstanding, of what we don't know about all those happy couples in black and white. A kiss, as a beginning and above all as an end. Black and white as an impossible dialogue, man and woman as the endless attempt at happiness which may only end up in grey.

Later, many other installations, pieces, in which intangible things are embodied in images, in the wills-o'-the-wisp of suspected presences, like in "Dress of Light", 1999. The projection of light onto a floating dress. The light isn't visible; we can only see it when it falls across a surface. And thus the empty dress of any body floats before us and disappears on a carousel that looks a lot like the merry-go-round of life. In 1985, she made a piece that has become a classic in the history of video art, *TV Weave*, an installation of televisions projecting only light on a single channel – hardly anything, a series of ripples of interference, a scratchy sound that floods the darkened room, lit only by the colored lights, endlessly oscillating and changing, emitted by the screens. There's nothing to see, only the light speaks to us from the other side of the screens. We don't understand what they're telling us, but we know that they're saying something just as decipherable as the hundreds of black and white kisses in *The End*.

And so the pieces and exhibitions follow one another, *Sincronías*, *Frequencies*, *Light Years*... Apparently because Balcells confesses that "In truth, I have only made one work in my whole life." In fact, she has done what artists do: try to understand the world, understand themselves. "I am also a spectator of my own works. I am also surprised when I see them." That surprise is simply seeing oneself from the outside, like in a dream. Sometimes the dreams are what's real.

Repetition is one of the basic techniques of learning and of the futile attempt to achieve perfection. That's how we learn to read: the "m" and the "a", make "ma", repeating. That's how we learn to love, breaking our hearts into little pieces over and over again. That's how we learn to live, waking up every day, after every night, trying again after every failure, every mistake. You've heard it before: fail again, fail better. And we do; each time we fail better, with more style, with more success. And once we reach that perfect failure, we've finally found what we're looking for. That's the final encounter, success, the way out of the loop.

Eugènia asserts that, "All wisdoms coincide, and that means unity and mystery." And we come to the point where we understand the whole story, including our own. Now we see that science and magic are a single thing, that the colors form their own and different alphabet, which is at the origins of life, an alphabet made up of 118 elements. Periodic elements that make up everything, all the shadows and all the lights.

Maybe one day we'll come to understand the hidden message in the waves of color on the monitors in *TV Weave*. In the journey to the origins represented by the exhibition *Light Years*, Balcells finds the crux of her continuance in life, and in art, which, like science and magic, like mystery, are one and the same. That journey to the zero point of life, to the past of our world and to the future of the universe, can only be made by riding the wild horse of color.

Everything is a beginning in discovery, because it has no end. Like every body, like every mind. Like life itself. Eugènia Balcells has entered that tunnel full of light that is the mystery of knowledge, and in the pages of this book many more voices, many more letters, words, feelings guide us along that path. Just follow the yellow brick road.

Biographies

Toni Pou (el Masnou, 1977) holds a degree in Physics from the University of Barcelona and is a scientific writer. He is the author of a book about a scientific expedition to the Arctic, *On el dia dorm amb els ulls oberts* [Where the Day Sleeps with Its Eyes Open] (Empúries, 2011; Anagrama, 2013), which was awarded the Godó Prize for Reporting and the Prisma Art Prize for the best popular science book published in Spain, and the novel *Si un dit assenyala la lluna* [If a Finger is Pointing at the Moon] (Anagrama, 2021), a moving defense of imagination and creativity applied to the sciences. He writes about science for the newspaper *Ara* in Barcelona and collaborates with various radio and television programs. A university professor of scientific communication, he is the co-founder of the company Eduscopi, dedicated to the dissemination of science, and curated the exhibition *The Arctic Is Breaking Up*, produced by CosmoCaixa, Barcelona's science museum.

Jordi Balló (Figueres, 1954) is a professor of Film Iconography at Pompeu Fabra University in Barcelona. He has been dean of the Faculty of Communication (2018-2021) and director of the Department of Communication (2021) at UPF, and has been a visiting professor at Stanford University (California, 2008). He is the author and co-author of the books *La llavor immortal. Els arguments universals en el cinema* [The Immortal Seed. Universal Plots in Cinema] (1995), *Imatges del silenci* Images of Silence (2000), *Jo ja he estat aquí: ficcions de la repetició* [I've Been Here Before: Fictions of Repetition] (2005, Serra d'Or Award), *El món, un escenari. Shakespeare: el guionista invisible* [All the World's a Stage, Shakespeare: The Invisible Screenwriter] (2015) and *Motivos visuales del cine* [Visual Motifs from Film] (2016). He was the conceptual author of the supplement "Cultura/s" for the newspaper *La Vanguardia* (2002 and 2014). He has also been the director of exhibitions at the Center for Contemporary Culture in Barcelona (1998-2011), where he has curated exhibitions such as *El segle del cinema* [A Century of Film], *Món TV* [TV World], *La ciutat dels cineastes* [The City of Filmmakers], *Erice/Kiarostami, Hammershøi and Dreyer, The Complete Letters: Filmed Correspondence*, and *Pasolini Roma*, where he explored the boundaries between the movie theater and the museum space. He directs the Master in Creative Documentary at Pompeu Fabra University, from where he has promoted films such as *La plaga* [The Plague] and *Sis dies corrents* [The Odd-Job Men], by Neus Ballús; *El cielo gira* [The Sky Turns], by Mercedes Alvarez; *Cravan vs. Cravan* and *La leyenda del tiempo* [The Legend of Time] by Isaki Lacuesta; *Aguaviva*, by Ariadna Pujol; *En construcción* [Under Construction], by José Luis Guerín, or *Mones com la Becky* [Monkeys like Becky] and *De nens* [For Kids], by Joaquim Jordà, among others.

Priyamvada Natarajan (Coimbatore, India, 1969) is a professor in the departments of Astronomy and Physics at Yale University. She is noted for her work probing the nature of dark matter and dark energy, using gravitational lensing, and for developing models that describe the assembly and growth histories of black holes in the universe. She was the first woman in Astrophysics to be elected a fellow at Trinity College. Natarajan's research work and original contributions to astrophysics have been recognized with many awards and honors. She is an elected fellow of the Royal Astronomical Society, the American Physical Society and the Explorers Club. In addition to her faculty position at Yale, Natarajan holds the Sophie and Tycho Brahe Professorship at the Niels Bohr Institute of the University of Copenhagen, Denmark and an honorary professorship for life at the University of Delhi, India. She is an Affiliate at the Black Hole Initiative at Harvard University and an Associate Member of the Center for Computational Astrophysics at the Flatiron Institute in New York. She is also the director of the Franke Program in Science and the Humanities at Yale, which fosters mutual understanding and collaborative teaching among diverse scientific and humanistic disciplines.She is the author of the critically acclaimed book *Mapping the Heavens: The Radical Scientific Ideas That Reveal the Cosmos* (2016).

Amandine Beyer (Aix-en-Provence, 1974) is an acclaimed artist in the interpretation of Baroque violin music. Her recording of J.S. Bach's sonatas and Partitas was received by critics as a revelation (Diapason d'or, Choc de l'année from *Classica*, Editor's Choice from Gramophone, the Académie Charles Cros Award, listed as "Exceptional" by *Scherzo*, among other accolades). Her work on this material continues with *Partita 2*, choreographed and danced by Anne Teresa de Keersmaeker and Boris Charmatz. A regular at the most important festivals in the world (Théâtre du Châtelet, Sablé Festival, Innsbruck Festwochen, Vienna Konzerthaus, etc.), her musical activities are divided between the different groups in which she plays: Les Cornets Noirs, duets with Pierre Hantaï, Kristian Bezuidenhout, Laurence Beyer and her own group, Gli Incogniti, with whom she has recorded various albums (Apothéoses by F. Couperin, Concerti Grossi by Corelli, Vivaldi's Four Seasons, Nicola Matteis, etc.). Those albums have received unanimous endorsement from international critics. She has been a teacher at the ESMAE in Porto (Portugal) and has given master classes all over the world: France, Italy, Spain, Brazil, Taiwan, the United States, and Canada. Since 2010, she has been a teacher of Baroque violin at the Schola Cantorum Basiliensis in Switzerland.

Federico Mayor Zaragoza (Barcelona, 1934) holds a PhD in Pharmacy from the Complutense University of Madrid and was a professor of Biochemistry at the Faculty of Pharmacy of the University of Granada, where he served as university rector. He launched the National Plan for the Prevention of Intellectual Disabilities. He is co-founder of the Severo Ochoa Center for Molecular Biology. Among other political responsibilities, he has held the positions of Undersecretary of Education and Science for the Spanish Government, Member of the Spanish Parliament, Counselor to the President of the Government, Adolfo Suárez, Minister of Education and Science, and Member of the European Parliament. In 1987, he was elected Director General of UNESCO and was re-elected in 1993 for a second term. He has served as president of the ERCEG (European Research Council Expert Group) for the "knowledge-based economy"; co-chairman of the High Level Group for the UN Alliance of Civilizations, and president of the Initiative for Science in Europe (ISE). He was president of the International Commission Against the Death Penalty, from its creation until 2017. He is director of the Scientific Council for the Ramón Areces Foundation, president of the Foundation Cultura de Paz and president of the Spanish Association for the Advancement of Science.

Simon McBurney (Cambridge, 1957) is a multi-Olivier Award-winning, Tony and SAG Award-nominated actor, writer, director and one of Europe's most original theatre makers. He is co-founder of Complicité, "Britain's most innovative theatre company" (*New Statesman*). His directing credits with Complicité include *The Encounter, Ungeduld des Herzens (Beware of Pity)*, a co-production with the Schaubühne, Berlin, *The Master and Margarita, Shunkin, A Disappearing Number, Measure for Measure, A Minute Too Late, The Elephant Vanishes, Strange Poetry* and *The Street of Crocodiles*. Other directing credits include *The Kid Stays in the Picture, All My Sons* on Broadway and *The Resistible Rise of Arturo Ui* with Al Pacino in New York. Opera credits include *The Rake's Progress* produced by Dutch National Opera and Festival d'Aix-en-Provence, *The Magic Flute* and *A Dog's Heart* both produced by Dutch National Opera and English National Opera in collaboration with Complicité. He is a prolific film, television and radio actor and has appeared in: *The Theory of Everything, Magic in the Moonlight, Tinker Tailor Soldier Spy, Harry Potter and the Deathly Hallows: Part 1*, and most recently *The Conjuring 2, Mission: Impossible - Rogue Nation, Allied, Carnival Row*, and *The Loudest Voice*.

Eulàlia Bosch (Barcelona, 1949) is a professor of philosophy. She is a founding member of the Research Institute for the Teaching of Philosophy (IREF), which she directed between 1987-1994. In 1995 she created the Department of Education at the Barcelona Museum of Contemporary Art, which she directed until 1997. She has curated exhibitions of contemporary art, including *Criatures misterioses* [Mysterious Creatures] (1992), *La ciutat de les paraules* [The City of Words] (1998), *Te mando este rojo cadmio* [I Send You This Cadmium Red] (2000), *Oteiza* (2000), *De l'Església a l'Auditori* [*From the Church to the Concert Hall* (Torroella de Montgrí, 2020), and *Només d'Anada* [One Way] (2020-2022), among others. Since 1996, she has worked with Eugènia Balcells and was responsible for organizing her exhibitions *Seeing the Light* (1996), *Frequencies* (2009) and *Light Years* (2012). She has always maintained her teaching activity, while also working as an editor, and she has published numerous articles and books, including *The Pleasure of Beholding: The Visitor's Museum* (Actar 1998), *Education and Everyday Life* (Hawker Brownlow 2003) and *Un lloc anomenat escola* [A Place Called School] (Graó 2009), all of which have been translated into several languages.

Sally Potter (London, 1949) made her first 8 mm film at age fourteen. She has since written and directed nine feature films, as well as many short films (including *Thriller* and *Play*) and a television series, and has directed opera (*Carmen* for the ENO in 2007) and other live performances. Her background is in choreography, music, performance art and experimental film. *Orlando* (1992), Sally Potter's bold adaptation of Virginia Woolf's classic novel, first brought her work to a wider audience. It was followed by *The Tango Lesson* (1996), *The Man Who Cried* (2000), *Yes* (2004), *Rage* (2009) and *Ginger & Rosa* (2012), and *The Party* (2017). Her latest film, *The Roads Not Taken* premiered at the Berlin Film Festival in 2020. Sally Potter has worked with many of the most notable cinema actors of our time, and her films have won over 40 international awards. She has had full career retrospectives of her film and video work at the BFI Southbank, London, MoMA, New York, and the Cinematheque, Madrid. Her book *Naked Cinema: Working with Actors* was published by Faber & Faber in March 2014. Sally Potter co-founded her production company Adventure Pictures with producer Christopher Sheppard.

Roberto Ontañón Peredo (Santander, 1965) is director of the Museum of Prehistory and Archaeology and the Prehistoric Caves of Cantabria. He holds a PhD in Prehistory and Archeology from the University of Cantabria (2000) and rounded out his studies in Paris (Universités Paris Ouest – Panthéon-Sorbonne – CNRS) (2001-2003). He collaborates with the French Ministry of Culture through several organizations dedicated to the investigation and conservation of caves containing rock art. He is also an adviser on rock art for UNESCO. A researcher at the International Institute of Prehistoric Research of Cantabria, he has been president of the Prehistoric Art Commission of the Union Internationale des Sciences Préhistoriques et Protohistoriques. His research focuses on prehistoric archeology and Paleolithic and post-Paleolithic cave art. Since 1995, he has co-directed the research project in the archaeological zone of La Garma. He is the author and editor of more than 250 scientific and popular articles. He has given lectures in Spain, France and Japan. He has been the curator of various exhibitions, notably *Iberian Picasso*, organized by the Centro Botín and the Musée National Picasso-Paris. Among other recognitions, he received the National Award for Archeology and Paleontology from the PALARQ Foundation in 2021.

María Muñoz (Valencia, 1963) was born to parents from Chera (Guadalajara) and Panticosa (Aragonese Pyrenees) and grew up in Valencia where she studied music and practiced competitive athletics. That was also where she started studying dance. Later she traveled to Amsterdam and Barcelona to continue her training. In 2005, she earned a degree in Dance in the specialty of Choreography and Performance techniques. She had her first professional experience in 1982, in the show *Era*, produced by the Japanese company Shusaku & Dormu Dance Theater, based in the Netherlands. Her career in creation for the stage began in 1985 with the formation of the group La Dux, together with Maria Antònia Oliver. In 1988, she collaborated with Pep Ramis to create the solo *Cuarto trastero* and in 1989 they formed the group Mal Pelo. Since then, the two have continued working together on creation and direction. Within the group she works as a researcher and a teacher of movement. She also edits and promotes the creation of texts to be performed on the stage. She is currently a choreographer, dancer and co-director of the creation center L'animal a l'esquena in Celrà, Girona.

Marc Balcells (Barcelona, 1958) is an astronomer. He earned his PhD from the University of Wisconsin–Madison (USA). He has carried out research in the United States, the Netherlands and Spain, and has published many research papers. He lives in the Canary Islands, where he is director of the Isaac Newton group of telescopes in La Palma. His research focuses on the formation, growth and transformation of galaxies, and he has used telescopes in the Canary Islands, Chile and Hawaii, as well as the Hubble Space Telescope. He promoted the development of world-leading infrared instrumentation for the Gran Telescopio Canarias. He regularly uses supercomputers to perform experiments on the role of mergers in galaxy morphogenesis.

Joaquim Sales (Barcelona, 1946) was a professor of Inorganic Chemistry at the University of Barcelona until 2016. At that same university, he began studies on organometallic compounds of transition elements and their catalytic applications. From 1995 onwards he became interested in the field of quantitative structure-property relationship (QSPR). He has been president of the Catalan Society of Physical, Chemical and Mathematical Sciences. In the field of the history of chemistry, he has published several articles on the periodic table, Linus Pauling, Edward Frankland, Josep Pascual and Enrique Moles. He served as the curator of the exhibition *Homenatge a Mendeléiev* [Tribute to Mendeleev] (CRAI of Physics and Chemistry-UB, 2007), and author of *La Química a la Universitat de Barcelona* [Chemistry at the University of Barcelona] (Edicions UB, 2011) and *Enrique Moles. Una biografia científica i política* [Enrique Moles. A Scientific and Political Biography] (Edicions UB and Editorial CSIC, 2021).

Rodolfo Häsler (Santiago de Cuba, 1958) is a poet and translator into Spanish from German, French and Catalan. He has lived in Barcelona since he was a child, and he is the son of the German-Swiss painter Rudolf Häsler. As a result, painting and the visual medium have had a profound effect on his writing. He has published 11 books of poetry; the last three are *Lengua de lobo* [Wolf's Tongue], Claudio Rodríguez International Poetry Prize (Ediciones Hiperión, 2019), *Cuaderno de Beirut* [Beirut Notebook] (Polibea, 2020) and *Hospital de cigüeñas* [Stork Hospital] (Libros de la hospitalidad, 2021). He is currently working on what will be his future book, *El tranvía verde de Alejandría* [The Green Streetcar of Alexandria].

Jaume Bertranpetit (Camprodon, 1952) is a professor of Biology at Pompeu Fabra University (Barcelona). He is the founder of the Evolutionary Biology Unit in the Department of Experimental and Health Sciences at that university as well as the Institute of Evolutionary Biology (CSIC-UPF). He is a former professor at the University of Barcelona and a disciple of Luigi Luca Cavalli-Sforza at Stanford University in California.

His fields of research are varied, but they all center on the study and understanding of the diversity of the human genome. He has published more than 300 scientific papers. He is a member of the boards of various scientific organizations (ICHN, SCB, ICA), he has been dean of the Faculty of Health and Life Sciences at UPF (1998-99), vice-dean for Science Policy at UPF (1999-2001) and director of the National Genotyping Center (1994-2010), and he was director of ICREA (Catalan Institution for Research and Advanced Studies) from 2007 to 2015. He is also a member of the Institut d'Estudis Catalans, the Royal Academy of Sciences and Arts of Barcelona, and the European Academy. Among other awards, he has received the Distinction in University Research from the General Directorate of Research, Generalitat de Catalunya; the Narcís Monturiol Medal for Research, and an honorary doctorate from Paul Sabatier University (Toulouse, 2018).

Cesc Gelabert (Barcelona, 1953) is a dancer, choreographer and educator. He began his dance studies with Anna Maleras. He studied architecture, and in 1972 he created his first choreography. In 1985, together with Lydia Azzopardi, he formed the Gelabert Azzopardi Dance Company, which has been in residence at the Teatre Lliure in Barcelona and the Hebbel Theater in Berlin. He has collaborated with first-rate companies and artists both nationally and internationally. He has received various awards, including the National Dance Award of Catalonia, the Gold Medal of Merit in the Fine Arts, the Gold Medal of the City of Barcelona, the National Dance Award from the Ministry of Culture, the Max Awards for the Performing Arts, the 2004 Herald Angel Award, and DAAD Berlin.

Sunetra Gupta (Calcutta, 1965) is an acclaimed novelist, essayist and scientist. Her fifth novel, *So Good in Black*, was published in February 2009. She was named as the winner of the 2009 Royal Society Rosalind Franklin Award for her scientific achievements. Sunetra, who lives in Oxford with her husband and two daughters, is Professor of Theoretical Epidemiology at Oxford University's Department of Zoology, having graduated in 1987 from Princeton University and received her PhD from the University of London in 1992. Sunetra wrote her first works of fiction in Bengali. She is an accomplished translator of the poetry of Rabindranath Tagore.

Stephen M. Noonan (Saint Louis, Missouri, 1961) is the founding principal of The Maxine Greene High School for Imaginative Inquiry (MGHS), formerly the High School for Arts, Imagination and Inquiry, located at the Martin Luther King, Jr. Educational Campus in Manhattan. The school was created in 2004 through a partnership between Maxine Greene, The New York City Department of Education (NYCDOE) and Lincoln Center Education (LCE). The school opened in September, 2005. It is the mission of MGHS to provide a learning community in which deep engagement with works of art enhances the imaginative capacity leading to the intellectual rigor of all students. This approach to inquiry-based learning is rooted in the philosophy of Dr. Maxine Greene, which fosters a sense of self-worth, curiosity and empathy. MGHS empowers this diverse community to work towards a more just, humane and vibrant world. In his role as principal, Mr. Noonan works to build strong, collaborative relationships with students and parents, while supporting a capable and diverse school faculty and staff. His professional philosophy is driven by the belief that by providing an individualized learning environment with a curriculum that is engaging and inquiry-based, all students will embrace the habits of lifelong learning. In 2009, the school joined the City University of New York (CUNY) School Support Organization. In 2015, the school became part of the Affinity Schools superintendency.

Marta Llorente Díaz (Girona, 1957) is an architect, a PhD, and a professor at the Barcelona School of Architecture, where she teaches courses in the theory of art and architecture, and the anthropology of the city, and leads a reading and writing workshop. She teaches Master's courses on literature and architecture (MBArch) and on space and gender (iiEDG, Inter-University Women and Gender Studies Institute). She directs the consolidated research group "Architecture, City and Culture" (ACC), which engages in research projects focusing on the contemporary city. She is a researcher at the iiEDG and the Eugenio Trías Center for Philosophical Studies. She has coordinated the research publications *Topología del espacio urbano* [Topology of Urban Space] (2014) and *Espacios frágiles en la ciudad contemporánea* [Fragile Spaces in the Contemporary City] (2019). She is the author of *El Saber de la arquitectura y de las artes* [Knowledge in Architecture and the Arts] (2000); *Susana Solano. Projectes* [Susana Solano: Projects] (2007); *La ciudad: inscripción y huella. Escenas y paisajes de la ciudad construida y habitada* [The City: Inscription and Trace. Scenes and Landscapes from the Built and Inhabited City] (2010); *La ciudad. Huellas en el espacio habitado* [The City: Traces in Inhabited Space] (2015); and *Construir bajo el cielo: un ensayo sobre la luz* [Building Under the Sky: An Essay on Light] (2020).

David Jou i Mirabent (Sitges, 1953) was a professor of Condensed Matter Physics at the Autonomous University of Barcelona until his retirement in 2018. He is a researcher in non-equilibrium thermodynamics, a subject about which he has published some 300 research papers and six books. He is also the author of an extensive body of work in the form of poetry and essays, which acts as a bridge between science, humanities and religion. Some of the titles include *Les escriptures de l'univers* [The Writings of the Universe] (poems about science), *La poesia de l'univers* [The Poems of the Universe] (science and mysticism), *Cerebro y universo: dos cosmologías,* [Brain and Universe: Two Cosmologies], *Reescribiendo el Génesis* [Rewriting Genesis], and *Introducción al mundo cuántico: de la danza de las partículas a las semillas de las galaxias* [Introduction to the Quantum World: From the Dance of Particles to the Seeds of Galaxies].

Roald Hoffmann (Złoczów, Poland, 1937) arrived in the U. S. in 1949, having survived World War II, and studied chemistry at Columbia and Harvard Universities (Ph.D. 1962). Since 1965, he has been at Cornell University, now as the Frank H. T. Rhodes Professor of Humane Letters, Emeritus. He has received many of the honors of his profession, including the 1981 Nobel Prize in Chemistry (shared with Kenichi Fukui). At Cornell, Hoffmann taught introductory chemistry half of his time. Notable also is his reaching out to the general public; he was the presenter, for example, of a television course in chemistry titled "The World of Chemistry," shown widely since 1990. Every first Sunday of the month, the Cornelia Street Café in Greenwich Village presented a cabaret evening, *Entertaining Science*, run by Hoffmann and Dave Sulzer, and featuring readings, music, dialogues between people or with audience, pieces of theater, magic, poetry, arguments on controversial issues – all having to do in one way or another with science. As a writer, Hoffmann has carved out a land between science, poetry, and philosophy, through many essays, five non-fiction books, three plays and seven published collections of poetry, including bilingual Spanish-English and Russian-English editions published in Madrid and Moscow.

F. Xavier Bosch José (Barcelona, 1947) earned a degree in Medicine from the University of Barcelona and specialized in Oncology at the Hospital de Sant Pau and in Public Health at UC Berkeley and UCLA. With a focus on cancer epidemiology and viral cancers, he worked for 10 years at the International Agency for Research on Cancer (IARC/WHO) in France, where he directed studies on human papillomavirus (HPV) and the origin of cancers in humans. In 1999 he returned to Barcelona, where he created and directed a research group in cancer epidemiology at the Catalan Institute of Oncology (ICO). The author of more than 500 international publications, he has received numerous awards, including an honorary doctorate from the University of Barcelona, the Maurice Hilleman Award for research on vaccines, the Rei Jaume I Prize and the Carlos IV Prize in the areas of preventive medicine and public health.

Amarjit Chandan (Nairobi, 1946) is a Punjabi writer, editor, translator and activist. He has written eight collections of poetry and five collections of essays in Punjabi. He has been called "the global face of modern Punjabi poetry". He has published over 25 books of poetry and essays, translated into many languages. He has edited over 15 books of poetry and essays. His two bilingual collections are *Sonata for Four Hands*, (Arc 2010), prefaced by John Berger, and *The Parrot, the Horse & the Man* (Arc 2017). He was one of 10 British poets selected by Andrew Motion, the Poet Laureate, on National Poetry Day in 2001. He has edited about 30 anthologies of Indian and world poetry and translated into Punjabi fiction by Brecht, Neruda, Ritsos, Hikmet, Cardenal, Martin Carter and John Berger, among others. Chandan formed a long-term association with John Berger. For Berger's 90th birthday in 2016, he co-edited *A Jar of Wild Flowers: Essays in Celebration of John Berger* and an anthology of poems by 90 poets *The Long White Thread of Words* (Smokestack).

Carlota Subirós (Barcelona, 1974) is a theater director, playwright and translator, born in Barcelona. She holds a degree in Stage Direction and Dramaturgy from the Institut del Teatre, for which she earned the Extraordinary Award, and in Italian Language and Literature from the University of Barcelona, for which she received the Honor Award. Her most recent work as a director includes the first stage adaptation of Doris Lessing's novel *The Golden Notebook*, as well as adaptations of works by John Berger, Mercè Rodoreda and Lewis Carroll. She has directed plays by authors such as Lluïsa Cunillé, Wallace Shawn, Tennessee Williams, Ödön von Horváth, Àngel Guimerà, Maxim Gorky or Antón Chéjov, apart from her own productions dealing with activism and politics. She is the author of two texts for family audiences, *La semilla del fuego* [The Seed of Fire] and the cantata *Un mundo entre dos tierras* [A World Between Two Lands].

Jean-Louis Froment (France, 1944) is a professor at the Bordeaux School of Fine Arts and the author of several books of poetry. In 1973, he founded the CAPC Musée de Bordeaux, which he led until 1996. Under his direction, the CAPC Musée developed a program that incorporates contemporary art, architecture, design and fashion. He also created an education department to build bridges between important intellectuals from around the world and the museum's audience. His exhibition projects, focusing on contemporary art, have traveled around Europe, Japan, America and Morocco and have been presented several times at the Venice Biennale. His voice was heard in Barcelona (1986-1994) as an advisor for the programming of the Fundació MACBA. Since 2018, he has been director of the artistic program Culture Chanel, associated with the image of Maison Chanel.

Salimata Wade (Dakar, 1951) has been a professor and researcher in the Department of Geography at Cheikh Anta Diop University in Dakar since 1996. The areas in which she participates in public intervention include education, training, research and cultural transmission.

In attempting to address the many challenges involved in feeding a population in the throes of a demographic and urban explosion, especially given the current context of climate change and various crises, Salimata Wade seeks to articulate issues related to security, governance, education, the market and food heritage. To participate in a more practical way in experimenting with solutions, in 2006, as a cook, Salimata Wade laid the foundations for La Compagnie du Bien Manger, which she created and has directed ever since. Between 2012 and 2017, Wade received the Harubuntu Prize and the three-year Ashoka Scholarship. In 2008, she founded Plus Value Culture, of which she is the General Secretary. The initiative's goal is to respond to the aim of providing better food for an ever-increasing number of people. In collaboration with the Micro Nutriment Initiative and the Polytechnical School at the University of Dakar, the association designed and locally produced the first machine to iodize salt in West Africa

Jorge Luis Ventocilla (Panama,1955) is the son of Peruvian exiles. He lived in Lima from the time he was 6 months old until he was 18 years old, when he moved to Panama to study biology (zoology) at the University of Panama. He holds a Master's degree in Wildlife Management (National University of Costa Rica). From 1980 to 2005, he worked at the Smithsonian Tropical Research Institute. His professional work has focused on environmental education and communication on issues related to the environment, indigenous peoples, and the interpretation of scientific projects in tropical biology. He is the author and co-author of books and articles, both technical and for the general public. He is a member of the Steering Committee for the Justo Arosemena Center for Latin American Studies (CELA). In 2005, the National Environmental Authority of Panama awarded him the first National Award for Environmental Excellence.

Maria Nadotti (Turin, 1949) is a journalist, essayist, editorial consultant and translator, and writes about theater, cinema, art and culture for Italian and foreign newspapers. She is the author of several essays, including *Silence = Death: The USA in the time of AIDS* (1994); *Cassandra non abita più qui* [Cassandra Doesn't Live Here Anymore] (1996); *Sesso & Genere* [Sex and Gender] (1996 and 2022); *Trasporti e traslochi. Raccontare John Berger* [Transport and Transfer: Explaining John Berger] (2014); *Necrologhi. Pamphlet sull'arte di consumare* [Obituaries: Pamphlet on the Art of Consumption] (2015). The books she has edited include *Off Screen: Women and Film in Italy* (1988); *Il cinico non è adatto a questo mestiere: Conversazioni sul buon giornalismo* [No Job for Cynics: Conversations on Good Journalism] (2000); *La speranza, nel frattempo. Una conversazione tra Arundhati Roy, John Berger e Maria Nadotti* [*Hope, Meanwhile. A Conversation between Arundhati Roy, John Berger and Maria Nadotti*] (2010); *Riga 32 - John Berger* [Row 32 - John Berger] (2012). A lover of the theater, in 1986 she organized a US tour for the Nobel Prize-winning actor and playwright Dario Fo and his wife, Franca Rame, and between 2002 and 2003 she conceived and planned a long tour for Pippo Delbono and his company in Palestine. In 2000, she adapted *Daughters of Ishmael in Wind and Storm*, by the Algerian writer Assia Djebar, for the Teatro di Roma, revived in 2009 at the Miller Theater in New York City. She is currently working on the stage adaptation of *To Be In A Time Of War*, by the Lebanese writer and painter Etel Adnan.

Jaume Casals (Barcelona, 1958) is a professor of Philosophy at Pompeu Fabra University, where he was dean from 2013 and 2021. He is a member of the Institut d'Estudis Catalans and earned a Dean's Medal from the Bloomberg School of Public Health (Johns Hopkins University). His lines of research include the Greek origins of modern and contemporary thought, as well as philosophy and literature. He has translated and edited works by Montaigne, Montesquieu, Berkeley and Bergson and is the author of some 50 academic articles and several books, including ¿Qué sé yo? La filosofía de Montaigne [What Do I Know? Montaigne's Philosophy] (Arpa, 2018), El aprendizaje de la muerte en la historia de las ideas [Learning about Death in the History of Ideas] (Ediciones UDP, 2009), El pou de la paraula [The Well of Words] (Edicions 62, 1996) and L'experiment d'Aristòtil [The Aristotle Experiment] (Edicions 62, 1992).

Holly Fairbank is the co-founder and Executive Director of The Maxine Greene Institute for Aesthetic Education and Social Imagination. She teaches courses in Arts & Aesthetic Education at Hunter College (CUNY) and Borough of Manhattan Community College (CUNY) and has taught similar courses at St. John's University, Lehman College and Queens College. Ms. Fairbank received a BA from Sarah Lawrence College and an MA in Dance Education from New York University. She was a teaching artist and an Assistant Director at Lincoln Center Institute (LCI) from 1997-2010. Holly was the artistic director of Holly Fairbank & Dancers, from 1979-1989. She has written numerous articles related to aesthetic education and her book, Collection, Preservation and Dissemination of Minority Dance in China: An Anthropological Investigation of the 1980's has been translated into Chinese and published by University of Yunnan Press, China.

Irene Martín (Shanghai, 1941) received an M.A. degree in Art History from Southern Methodist University (Texas) in 1980. From June 1987 to July 1994, she was director of the Thyssen-Bornemisza Foundation museum at Villa Favorita, Lugano. She worked as Exhibitions Manager at the J. Paul Getty Museum, Los Angeles from July 1994 to May 1998. As her most recent employment, she was Assistant Director, Exhibition Programs at the Los Angeles County Museum of Art (LACMA) from June 1998 to June 2010 and served as Senior Exhibitions Officer at LACMA from July 2010 to January 2013. She is a founding member of the International Exhibitions Organizers group (IEO). She is now retired, living in Dallas, Texas, and doing volunteer work with the Dallas Museum of Art and the Crow Museum of Asian Art/University of Texas, Dallas.

Noni Benegas (Buenos Aires, 1974) lives in Madrid. Her best poems are collected in the book El ángel de lo súbito [The Angel of Suddenness] (FCE, Madrid, 2014). Burning Cartography (Host, Austin TX, 2007 and 2011) offers them in English and Animaux Sacrés (Al Manar, Séte 2013), in French. She has received the Platero prize from the Spanish book club at the UN in Geneva; the Miguel Hernández National Prize, the Vila de Martorell prize, the Rubén Darío prize and the Esquío prize. She is the author of the influential anthology Ellas tienen la Palabra. Dos décadas de Poesía Española [They Have the Floor: Two Decades of Spanish Poetry] (Ed. Hiperión, Madrid, 2008, 4ta. edición). Her introductory essay was reedited by Fondo de Cultura Económica in 2018 as Ellas tienen la palabra. Las mujeres y la escritura [They Have the Floor: Woman and Writing]. In 2019 she published Ellas resisten (Huerga&Fierro, Madrid) about women writers and artists. Her collection of poems Falla la noche is forthcoming. She was awarded the Yourcenar scholarship in 2009. She is a lecturer at national and foreign universities, and she is a patron of the Voix Vives festival in Toledo. Her poems have been translated into English, French, Portuguese, Greek, Chinese and Italian.

Santiago Alvarez (Panama, 1950) studied chemistry at the University of Barcelona (PhD 1980). He is now Emeritus Professor of Inorganic Chemistry at the University of Barcelona, where he has developed theoretical research on bonding, stereochemistry and magnetic properties of transition metal compounds, and on the application of continuous shape and symmetry measures to structural and stereochemical studies. He has been awarded, among others, the prize for research in Inorganic Chemistry from the Real Sociedad Española de Química in 2003, the Solvay prize for research in Chemical Science in 2003, the Catalan-Sabatier prize from the Société Chimique de France in 2012, and the Medal of the Real Sociedad Española de Química in 2020. Alvarez has also promoted encounters across the borders of science and humanities through essay articles, three books, and the biannual NoSIC (Not Strictly Inorganic Chemistry) meetings held between 2003 and 2020. He is a Fellow of the Royal Society of Chemistry, Corresponding Member of the Spanish Academy of Sciences, Member of the European Academy of Sciences, and a Lifetime Honorary Member of the Israel Chemical Society.

Thane Lund (Crosby, North Dakota, USA, 1986) is an artist based in Brooklyn, NY. His work has been exhibited in the United States, Ireland, Germany, China, and Japan. He has participated in numerous artist residencies including the Burren College of Art Alumni Residency in Ballyvaughan, Ireland; the Vermont Studio Center, Studio Kura in Itoshima, Japan; and the Varda Artist Residency in Sausalito, California. He holds a Bachelor of Fine Arts degree from Montana State University.

Rosa Olivares (Madrid, 1955) is a writer, editor, art critic and independent curator. Since 1985, she has collaborated in the press specialized in visual arts in Spain and in various other countries. She has taught courses and given lectures in Spain, Germany, Colombia, Mexico, Uruguay, Argentina and Chile, among other places. She has curated exhibitions in museums and art centers in Spain, France, Italy, Germany, Argentina, Mexico, Chile, Peru, Panama, and Costa Rica, etc. She has edited and directed magazines such as *Lapiz, Exit, ExitBook, Exit-Express, Flúor* and *Utopía*. In addition, she has written and edited dozens of publications. She directs the programming for the photography room at the Oscar Niemeyer Foundation in Avilés, Spain. She currently resides between Spain and Mexico, where she runs a bookstore specialized in contemporary art and urban culture.

Eugènia Balcells (Barcelona, 1943) holds a degree in Quantity Surveying. The daughter and granddaughter of architects and inventors, her daily contact with all kinds of ingenious installations related to vision and mathematics helped her begin to understand the fragile balance between the intangible and the material, between imagination and precision. In 1968, she moved to New York City and continued her artistic training at the University of Iowa where she obtained a Master of Arts in 1971. She lived between Barcelona and New York City until 1979, when she finally settled in the United States. As of 1988, she once again began residing alternately in both cities. She began her artistic career in the mid-1970s in the world of conceptual art, and she became one of the pioneers of experimental cinema and audiovisual art in Spain. Her first installations, films and videos were aligned with critical-sociological currents, dealing with issues related to the consumer society and the effects of mass media on popular culture.

Starting in the late 1970s, she incorporated the circle into some of her most significant pieces – on the one hand as a physical description of the movement of a camera located at the center of a space, and on the other as a formal conception used in an installation's compositional system. This strategy is used in the film *Fuga* [Fugue] (1979), in the video *Indian Circle* (winner of the Grand Prix at the 1ère Manifestation Internationale de Video in Montbeliard, 1982) and in the video installation *From the Center* (which won an award from the Visual Studies Workshop, in Rochester, N.Y. in 1983).

From 1981 to 1982, as a result of a series of contacts with various North American musicians, including Peter Van Riper, and continuing with the exploration of the relationship between images and sounds seen in her film *133* (1978-79), she created a series of pieces called *Sound Works*.

In her exploration of the limits of visual perception through the two video installations *Color Fields*, presented at the I National Video Festival at the Círculo de Bellas Artes in Madrid (1984) and *TV Weave*, presented at the Institute for Art and Urban Resources, PS1, New York (1985), she offered a new understanding of the electronic image. With *TV Weave*, she participated in the exhibition *Primera generación - Arte e Imagen en Movimiento* [First Generation - Art and Moving Images] at the Reina Sofía National Museum Art Center (2006-2007). The exhibition was structured around different ideas and pieces by the world's first video artists.

How human beings process their own existence, and the harmonies and dissonances in interpersonal relationships, are central themes in works such as *Descansan como en la casa materna I y II*, *En trànsit*, and *Sincronías* presented at the Reina Sofía National Museum Art Center in Madrid in 1995.

The interaction of the complementary functions of the brain's two hemispheres are the object of reflection in the installation *Transcending Limits* (part of the exhibition *Veure la llum* [Seeing the Light] at the MACBA in 1996), in which six translucent screens are crossed by two opposite projections, acting as a metaphor for the coexistence of the rational and intuitive worlds.

Balcells' work includes a large number of audiovisual installations. Sometimes she reflects explicitly on light, one of the interests that has structured her creative and teaching research.

Her latest work for public space was a garden of light, *Jardí de llum - Collserola* (2003), a permanent installation located in the Ciutat Meridiana metro station in Barcelona.

The interest in everyday objects, already present in her early works (*Supermercart*, *Clear Books*), is the foundation for an exploration of the human habitat through the objects belonging to each space and their symbolic and energetic essence. The exhibition *En el cor de les coses* [In the Heart of Things] is made up of five audiovisual installations that correspond to the essential spaces of a house (living room, kitchen, dining room, bedroom and bathroom) and was presented at Tinglado 2 in the Port of Tarragona in 1998.

Another work that uses everyday objects, *Roda do tempo* [Wheel of Time], is a multiple reflection on time: the cyclical time of nature, the dialectical time of history, subjective time, imaginary time, the non-existence of time, and the coexistence of all times. This installation was commissioned by the Centro Cultural Banco do Brasil, Rio de Janeiro, in 2001.

In some of her works she deals with the image and the role of women in culture and the need to honor our historical legacy (*Going through Languages*, *Álbum portátil*), and she portrays women undergoing a constant transformation, bathed in the waves of the sea as a metaphor for life in *Un espai propi* [A Space of One's Own] (2000). This piece, a tribute to Virginia Woolf, was commissioned by the Center for Contemporary Culture of Barcelona (CCCB). It explores the boundaries between the interior and exterior world.

As a welcome to the new millennium, the artist created *Brindis* [Toast] (1999) presented at the Alter Ego gallery in Barcelona and at the Casa de las Américas in Madrid. The piece invokes abundance, plenitude, happiness and celebration. It is a celebration of life that invites us to fill up our glasses and make a toast to our fondest desires.

Her installation *Anar-hi anant* [On and On] (2000), a synthetic work that uses everyday objects to generate a new poetic experience of perception, was part of the exhibition *El discreto encanto de la tecnología-artes en España* [The Discreet Charm of Technology-Arts in Spain] at the MEIAC Museum in Badajoz and at the ZKM Center for Art and Media, in Karlsruhe, Germany and was presented in July 2009 for the Fontana d'Or in the city of Girona.

In 2007, she began the artistic project *Frequencies* that centers on light and color as elements that distinguish perception – permitting it and limiting it at the same time. The piece aims to create nearly holographic images, in which different planes coexist in order to represent reality as a fabric of frequencies. The poetic character of the work is rooted in the encounter between the sciences and the arts, always so fragile and so powerful. This project was presented in September 2009 at Arts Santa Mònica in Barcelona and began touring at the Gas Natural-Fenosa Contemporary Art Museum in La Coruña (MACUF, 2010-2011), including *Homage to the Elements*, a version of the periodic table that includes the emission spectra of all the elements.

In the film *Rice Is Planted with Rice*, Balcells explores the educational impact of *Frequencies* at different schools in the city of Barcelona.

Universe and the film *Glimpsing the Universe* are two works created in 2012, presented for the first time at the Tabacalera in Madrid, together with *Frequencies* and the mural *Homage to the*

Elements, to form the exhibition *Light Years*, which later traveled to Mexico (Cenart, 2015 and Querétaro National Center for the Arts, 2016), Panama (MAC, 2017) and Barcelona (CosmoCaixa, 2018).

Her activities also include teaching courses at different schools and universities. In her workshops, she works with light as a creative essence, integrating different forms of knowledge such as philosophy, literature, poetry, and physics, among others, into her vision.

Eugènia Balcells' creative activity is currently focused on the project *Legacy*, forthcoming, which comprises multiple multimedia installations and graphic work, in which she reviews the relationship between reading and visual perception.

In 2021, she moved her studio from Brooklyn (NY) to Catalonia and began work on the new headquarters of the Eugènia Balcells Foundation in Castellar de la Selva, in the municipality of Quart, near Girona.

In February 2010, she was awarded the Medal of Merit in Fine Arts 2009 in the Gold category by H.M. the King. In 2010, on the occasion of her exhibition *Frequencies*, the National Council for Culture and the Arts awarded her the National Prize for Visual Arts from the Generalitat de Catalunya.

My Name is Univers

Author
Toni Pou

Edited by
Eulàlia Bosch

Fotography
Eugènia Balcells

Afterword
Rosa Olivares

Publisher
Fundació Eugènia Balcells
Actar Publishers

Translation
Angela Kay Bunning

Graphic Design
Ramon Prat Homs

Distribution
Actar Distribution Inc.

Barcelona
Roca i Batlle 2-4
08023 BARCELONA, Spain
T +34 933 282 183
eurosales@actar-d.com

New York
440 Park Avenue South,
17th Floor
NEW YORK, NY 10016, USA
T +1 2129662207
salesnewyork@actar-d.com

ISBN: 978-1-63840-077-6
PCN: Library of Congress Control
Number: 2022947234
Printed in Europe, 2022

It is possible to say with a fair amount of conviction that no book has ever been written by just one person. That is even more true in this case: not only was this book made possible by contributions from many people, it is literally made up of those contributions. Many people from places all over the planet had a hand in making it, working around the well-known difficulties associated with geographic distance and the ones brought about, unexpectedly, by the virus that has turned the world on its head in recent years. Without the efforts and the generosity of

Jordi Balló, Priyamvada Natarajan, Amandine Beyer,
Eugènia Balcells, Federico Mayor Zaragoza, Simon McBurney,
Eulàlia Bosch, Sally Potter, Roberto Ontañón, María Muñoz,
Marc Balcells, Joaquim Sales, Rodolfo Häsler,
Jaume Bertranpetit, Cesc Gelabert, Sunetra Gupta,
Stephen M. Noonan, Marta Llorente, David Jou
and
Roald Hoffmann, Xavier Bosch, Amarjit Chandan,
Carlota Subirós, Jean-Louis Froment, Salimata Wade,
Jorge Ventocilla, Maria Nadotti, Jaume Casals, Holly Fairbank,
Irene Martin, Noni Benegas, Santiago Álvarez, Thane Lund,

this book would not be in your hands.
One hundred eighteen periodic thanks to of all of them.

Toni Pou